Principles of Soil Microbiology

Principles of Soil Microbiology

Marcus Silva

STATES
ACADEMIC PRESS
www.statesacademicpress.com

Published by States Academic Press,
109 South 5th Street,
Brooklyn, NY 11249, USA
www.statesacademicpress.com

Principles of Soil Microbiology
Marcus Silva

International Standard Book Number: 978-1-63989-442-0 (Hardback)

Cataloging-in-Publication Data

Principles of soil microbiology / Marcus Silva.
p. cm.
Includes bibliographical references and index.
ISBN 978-1-63989-442-0
1. Soil microbiology. 2. Soil biology. 3. Soilborne plant diseases. I. Silva, Marcus.
QR111 .P75 2022
579.175 7--dc23

TABLE OF CONTENTS

PREFACE

The purpose of this book is to help students understand the fundamental concepts of this discipline. It is designed to motivate students to learn and prosper. I am grateful for the support of my colleagues. I would also like to acknowledge the encouragement of my family.

A microorganism is a microscopic organism that may exist in single-celled form or in colony of cells. The study of microorganisms that are present in the soil is known as soil microbiology. It is also concerned with the study of the functions of microorganisms and how they affect the properties of the soil. These microorganisms fix the nitrogen content in the soil and release oxygen in the atmosphere. Some of these microorganisms present in the soil are bacteria, fungi, protozoa, actinomycetes and algae. This area of study is majorly used in agriculture. It makes the minerals and nutrients present in soil available to the plants and help them produce hormones that initiate growth. This book unfolds the innovative aspects related to this field which will be crucial for the holistic understanding of the subject matter. It is a valuable compilation of topics, ranging from the basic to the most complex theories and principles in the field of soil microbiology. This book is an essential guide for both academicians and those who wish to pursue this discipline further.

A foreword for all the chapters is provided below:

Chapter – What is Soil Microbiology?

Soil microbiology is the branch of science that deals with the study of microorganisms in soil, their functions and their effects on soil properties. Soil microbes decompose and recycle nutrients and fixes nitrogen from the air for the growth of plants. This is an introductory chapter which will briefly introduce about soil microbiology.

Chapter – Basic Concepts of Soil Microbiology

Some of the basic concepts of soil microbiology include sustainable agriculture, soil organic matter, carbon and phosphorus cycle, nitrogen fixation, nitrification and denitrification, etc. This chapter has been carefully written to provide an easy understanding of the concepts related to soil microbiology.

Chapter – Soil Microorganisms

Soil microorganisms can be categorized into bacteria, actinomycetes, fungi, algae and protozoa. Each of these categories has different characteristics based upon the functions and properties of soil. This chapter closely examines all the aspects concerned with soil microorganisms to provide an extensive understanding of the subject.

Chapter - Soil Environments

Soil environment means the physical, biological and chemical factors that affect the microbes present in the soil. Soil habitat, soil ecology, microbial community, soil nitrogen dynamics, etc. are studied under its domain. The topics elaborated in this chapter will help in gaining a better perspective of soil environment.

Chapter - The Rhizosphere

The rhizosphere is the narrow region of soil around the roots where microbial biomass is affected by the presence of soil microbes, plant roots and their secretions. Some of the rhizosphere organisms are oomycetes, bacteria, pathogenic fungi, and human pathogens. All the topics associated with the rhizosphere have been carefully analyzed in this chapter.

Marcus Silva

What is Soil Microbiology? 1

- **Scope and Importance of Soil Microbiology**
- **Role of Soil Microorganisms**
- **Soil Microbes and Nutrient Recycling**
- **Plant-microbe Ecology**

Soil microbiology is the branch of science that deals with the study of microorganisms in soil, their functions and their effects on soil properties. Soil microbes decompose and recycle nutrients and fixes nitrogen from the air for the growth of plants. This is an introductory chapter which will briefly introduce about soil microbiology.

Soil microbiology is the scientific discipline that is concerned with the study of all biological aspects of the life that exist in the soil environment. Microorganisms in soil are important because they affect soil structure and fertility. Soil microorganisms can be classified as bacteria, actinomycetes, fungi, algae and protozoa. Each of these groups has characteristics that define them and their functions in soil. Soil also provides habitats colonized by a staggering variety of microorganisms. All these forms of life interact with one another and also interact with the soil to create continually changing conditions. This allows an on-going evolution of soil habitats.

Healthy soils contain numerous populations of organisms—from microorganisms to other organisms to large burrowing animals—that are essential for many of our biogeochemical cycles as well as for plant growth. The abundance of life that exists within our soils exceeds that of any other ecosystem. Aside from these obvious benefits, it is thought that soils hold a potential reserve of organisms that can provide antibiotics and other medically important functions.

The most prolific species found in soils are bacteria, which are most abundant around the roots of plants. Through the process of bacteria decomposing organic materials, nutrients that would be bound up in the soil are made available for plant, animal and human nourishment. Fungi are another important recycler of nutrients, in addition to protecting plants by consuming insects and organisms that prey on plants. While common in ponds and streams, algae are also common in soils, contributing to soil building and making it possible for plant growth. Many algae also fix nitrogen. Together, fungi and algae form lichens which are able to be productive while inhabiting harsher environments, even those with scarce water and nutrients. Lichens can also serve as indicators of environmental quality due to their ability to absorb trace inorganic and organic materials, regardless of toxicity.

Larger soil organisms include a wide variety of worms and arthropods, including nematodes and earthworms, which play a role in maintaining soil quality while also recycling nutrients. Nematodes are one of the world's most prolific species as predators to control the bacterial population as well as the number of protozoa and other organisms. Earthworms help to shape soil quality by loosening and aerating the soil while consuming plant and organic matter and converting it to humus. Earthworm "castings," the product of their digestion, are rich and less acidic than the surrounding soil, containing additional source nutrients of calcium, nitrogen, phosphate, and potassium.

Burrowing animals, such as mice, squirrels, rabbits and badgers, live part of the time in the soil as protection from predators or extreme weather. As predators themselves, they help control insect and arthropod populations within the soil. Their digging fertilizes and aerates the soil by mixing various subsurface materials. This also provides for an easier infiltration of water into the soil. In addition to mixing oxygen, their burrowing helps to bury carbon as well as seeds that can establish new plant growth.

Scope and Importance of Soil Microbiology

Living organisms both plant and animal types constitute an important component of soil. Though these organisms form only a fraction (less than one percent) of the total soil mass, but they play important role in supporting plant communities on the earth surface. While studying the scope and importance of soil microbiology, soil-plant-animal ecosystem as such must be taken into account. Therefore, the scope and importance of soil microbiology, can be understood in better way by studying aspects like:

- Soil as a living system.
- Soil microbes and plant growth.
- Soil microorganisms and soil structure.
- Organic matter decomposition.
- Humus formation.

- Biogeochemical cycling of elements.
- Soil microorganisms as bio-control agents.
- Soil microbes and seed germination.
- Biological N_2 fixation.
- Degradation of pesticides in soil.

Soil as a living system: Soil inhabit diverse group of living organisms, both micro flora (fungi, bacteria, algae and actinomycetes) and micro-fauna (protozoa, nematodes, earthworms, moles, ants). The density of living organisms in soil is very high i.e. as much as billions/gm of soil, usually density of organisms is less in cultivated soil than uncultivated/virgin land and population decreases with soil acidity. Top soil, the surface layer contains greater number of microorganisms because it is well supplied with Oxygen and nutrients. Lower layer/subsoil is depleted with Oxygen and nutrients hence it contains fewer organisms. Soil ecosystem comprises of organisms which are both, autotrophs (Algae, BOA) and heterotrophs (fungi, bacteria). Autotrophs use inorganic carbon from CO_2 and are "primary producers" of organic matter, whereas heterotrophs use organic carbon and are decomposers/consumers.

Soil microbes and plant growth: Microorganisms being minute and microscopic, they are universally present in soil, water and air. Besides supporting the growth of various biological systems, soil and soil microbes serve as a best medium for plant growth. Soil fauna & flora convert complex organic nutrients into simpler inorganic forms which are readily absorbed by the plant for growth. Further, they produce variety of substances like IAA, gibberellins, antibiotics etc. which directly or indirectly promote the plant growth.

Soil microbes and soil structure: Soil structure is dependent on stable aggregates of soil particles-Soil organisms play important role in soil aggregation. Constituents of soil are viz. organic matter, polysaccharides, lignins and gums, synthesized by soil microbes plays important role in cementing/binding of soil particles. Further, cells and mycelial strands of fungi and actinomycetes, Vormicasts from earthworm is also found to play important role in soil aggregation. Different soil microorganisms, having soil aggregation/soil binding properties are graded in the order as fungi > actinomycetes > gum producing bacteria > yeasts.

Examples are: Fungi like Rhizopus, Mucor, Chaetomium, Fusarium, Cladasporium, Rhizoctonia, Aspergillus, Trichoderma and Bacteria like Azofobacler, Rhizobium Bacillus and Xanlhomonas.

Soil microbes and organic matter decomposition: The organic matter serves not only as a source of food for microorganisms but also supplies energy for the vital processes of metabolism that are characteristics of living beings. Microorganisms such as fungi, actinomycetes, bacteria, protozoa etc. and macro organisms such as earthworms, termites, insects etc. plays important role in the process of decomposition of organic matter and release of plant nutrients in soil. Thus, organic matter added to the soil is converted by oxidative decomposition to simpler nutrients/substances for plant growth and the residue is transformed into humus. Organic matter/substances include cellulose, lignins and proteins (in cell wall of plants), glycogen (animal tissues), proteins and fats (plants, animals). Cellulose is degraded by bacteria, especially those of genus Cytophaga and other genera (Bacillus, Pseudomonas, Cellulomonas, and Vibrio Achromobacter) and fungal genera (Aspergillus, Penicilliun, Trichoderma, Chactomium, Curvularia). Lignins and proteins are partially digested by fungi, protozoa and nematodes. Proteins are degraded to individual amino acids mainly by fungi, actinomycetes and Clostridium. Under unaerobic conditions of waterlogged soils, methane are main carbon containing product which is produced by the bacterial genera (strict anaerobes) Methanococcus, Methanobacterium and Methanosardna.

Soil microbes and humus formation: Humus is the organic residue in the soil resulting from decomposition of plant and animal residues in soil, or it is the highly complex organic residual matter in soil which is not readily degraded by microorganism, or it is the soft brown/dark coloured amorphous substance composed of residual organic matter along with dead microorganisms.

Soil microbes and cycling of elements: Life on earth is dependent on cycling of elements from their organic/elemental state to inorganic compounds, then to organic compounds and back to their elemental states. The biogeochemical process through which organic compounds are broken down to inorganic compounds or their constituent elements is known "Mineralization", or microbial conversion of complex organic compounds into simple inorganic compounds & their constituent elements is known as mineralization.

Soil microbes plays important role in the biochemical cycling of elements in the biosphere where the essential elements (C, P, S, N & Iron etc.) undergo chemical transformations. Through the process of mineralization organic carbon, nitrogen, phosphorus, Sulphur, Iron etc. are made available for reuse by plants.

Soil microbes and biological N_2 fixation: Conversion of atmospheric nitrogen in to ammonia and nitrate by microorganisms is known as biological nitrogen fixation.

Fixation of atmospheric nitrogen is essential because of the reasons:

- Fixed nitrogen is lost through the process of nitrogen cycle through denitrification.
- Demand for fixed nitrogen by the biosphere always exceeds its availability.
- The amount of nitrogen fixed chemically and lightning process is very less (i.e. 0.5%) as compared to biologically fixed nitrogen.

- Nitrogenous fertilizers contribute only 25% of the total world requirement while biological nitrogen fixation contributes about 60% of the earth's fixed nitrogen.
- Manufacture of nitrogenous fertilizers by "Haber" process is costly and time consuming.

The numbers of soil microorganisms carry out the process of biological nitrogen fixation at normal atmospheric pressure (1 atmosphere) and temp (around 20 °C).

Two groups of microorganisms are involved in the process of BNF.

Non-symbiotic (Free Living) and B. Symbiotic (Associative):

- Non-symbiotic (Free Living): Depending upon the presence or absence of oxygen, non symbiotic N_2 fixation prokaryotic organisms may be aerobic heterotrophs (Azotobacter, Pseudomonas, Achromobacter) or aerobic autotrophs (Nostoc, Anabena, Calothrix, BGA) and anaerobic heterotrophs (Clostridium, Kelbsiella. Desulfovibrio) or anaerobic Autotrophs (Chlorobium, Chromnatium, Rhodospirillum, Meihanobacterium etc).
- Symbiotic (Associative): The organisms involved are Rhizobium, Bratfyrhizobium in legumes (Aerobic): Azospirillum (Grasses), Actinonycetes frantic(with Casuarinas, Alder).

Soil microbes as biocontrol agents: Several ecofriendly bioformulations of microbial origin are used in agriculture for the effective management of plant diseases, insect pests, weeds etc. eg: Trichoderma sp and Gleocladium sp are used for biological control of seed and soil borne diseases. Fungal genera Entomophthora, Beauveria, Metarrhizium and protozoa Maltesia grandis. Malameba locustiae etc are used in the management of insect pests. Nuclear polyhydrosis virus (NPV) is used for the control of Heliothis/ American boll worm. Bacteria like Bacillus thuringiensis, Pseudomonas are used in cotton against Angular leaf spot and boll worms.

Degradation of pesticides in soil by microorganisms: Soil receives different toxic chemicals in various forms and causes adverse effects on beneficial soil micro flora/micro fauna, plants, animals and human beings. Various microbes present in soil act as the scavengers of these harmful chemicals in soil. The pesticides/chemicals reaching the soil are acted upon by several physical, chemical and biological forces exerted by microbes in the soil and they are degraded into non-toxic substances and thereby minimize the damage caused by the pesticides to the ecosystem. For example, bacterial genera like Pseudomonas, Clostridium, Bacillus, Thiobacillus, Achromobacter etc. and fungal genera like Trichoderma, Penicillium, Aspergillus, Rhizopus, and Fusarium are playing important role in the degradation of the toxic chemicals/pesticides in soil.

Biodegradation of hydrocarbons: Natural hydrocarbons in soil like waxes, paraffin's, oils etc are degraded by fungi, bacteria and actinomycetes. E.g. ethane (C_2H_6) a paraffin hydrocarbon is metabolized and degraded by Mycobacteria, Nocardia, Streptomyces Pseudomonas, Flavobacterium and several fungi.

Role of Soil Microorganisms

In their natural environment, plants are part of a rich ecosystem including numerous and diverse microorganisms in the soil. It has been long recognized that some of these microbes, such as mycorrhizal fungi or nitrogen fixing symbiotic bacteria, play important roles in plant performance by improving mineral nutrition. However, the full range of microbes associated with plants and their potential to replace synthetic agricultural inputs has only recently started to be uncovered. In the last few years, a great progress has been made in the knowledge on composition of rhizospheric microbiomes and their dynamics. There is clear evidence that plants shape microbiome structures, most probably by root exudates, and also that bacteria have developed various adaptations to thrive in the rhizospheric niche. The mechanisms of these interactions and the processes driving the alterations in microbiomes are, however, largely unknown.

Interconnection of Plants with Soil Microbes

Although plant physiologists sometimes view soil as simply a source of nutrients to plants, it is actually a complex ecosystem hosting bacteria, fungi, protists, and animals. Plants exhibit a diverse array of interactions with these soil-dwelling organisms, which span the full range of ecological possibilities (competitive, exploitative, neutral, commensal, mutualistic). Throughout modern plant science, most interaction studies have focused on alleviating pathogenic effects such as herbivory and infection, or attenuating abiotic stress conditions. However, there has also been longstanding interest in characterizing the positive ecological interactions that promote plant growth. For instance, mycorrhizal fungi as well as the bacteria present in nodulated legumes were both recognized as root symbionts from the second half of 19th century. Already in the 1950s, crop seeds were coated with bacterial cultures (Azotobacter chroococcum or Bacillus megaterium) to improve growth and yield. By the 1980s many different bacterial strains, mainly Pseudomonas but also Azospirillum, had been described as having plant growth promoting effects. Since the 2000s, research focus has somewhat shifted away from individual microbial strains, and toward documenting the abundance and diversity of the root microbiome through metagenomics. Results from such sequencing studies have shown that the rhizospheric niche is a hotspot of ecological richness, with plant roots hosting an enormous array of microbial taxa. In the last few years, research has swung toward assembling rationally designed synthetic communities (SynComs) that comprise strains representing the dominant rhizospheric taxa, with the aim of re-capitulating favorable microbial functions under controlled experimental conditions. A major goal of this research field is to gain a mechanistic understanding of how soil microbes boost plant growth and defense, and then to use this knowledge to inform the optimal design of microbial communities tailored to carry out specific functions.

Microbial Traits and the Bioavailability of Nutrients for Plants

Three mechanisms are usually put forward to explain how microbial activity can boost plant growth: (1) Manipulating the hormonal signaling of plants (2) Repelling or outcompeting pathogenic microbial strains; and (3) Increasing the bioavailability of soil-borne nutrients. In natural ecosystems, most nutrients such as N, P, and S are bound in organic molecules and are therefore minimally bioavailable for plants. To access these nutrients, plants are dependent on the growth of soil microbes such as bacteria and fungi, which possess the metabolic machinery to depolymerize and mineralize organic forms of N, P, and S. The contents of these microbial cells are subsequently released, either through turnover and cell lysis, or via protozoic predation. This liberates inorganic N, P, and S forms into the soil, including ionic species such as ammonium, nitrate, phosphate, and sulfate that are the preferred nutrient forms for plants. In natural settings, these microbial nutrient transformations are key drivers of plant growth, and can sometimes be the rate-limiting step in ecosystem productivity.

Fertilization Practices and Environmental Sustainability

In most contemporary agricultural systems, macronutrients are provided through the application of mineral fertilizers. However, unsustainable fertilization practices are contributing to the large-scale alterations of Earth's biogeochemical cycles, through mechanisms such as soil degradation, waterway eutrophication, and greenhouse gas emissions. Furthermore, known reserves of phosphate rock are rapidly diminishing and predicted to be exhausted within a few decades, while N-fertilizer production via the energy-intensive Haber–Bosch process relies upon fossil fuels and thus exacerbates global warming and natural resource depletion. Due to the scale and severity of these fertilizer-induced problems, a current research priority is for agricultural science to develop alternative methods of sustaining plant nutrition with dramatically lower inputs of mineral fertilizers. One such possibility is to replace mineral fertilizers by organic inputs, and to supplement plants with specific root-associated microbes that depolymerize and mineralize the organic-bound nutrients. The logic of this idea is that organic inputs can be obtained more sustainably than mineral fertilizers, because myriad agricultural, industrial and municipal processes produce huge volumes of nutrient-rich "waste" that are currently disposed of, but could potentially be composted and applied as fertilizers.

Another factor is that organically bound nutrients are more stable in the soil compared to mineral fertilizers, and therefore less prone to leaching and volatilization. Bio-fertilizers are already used in organic farming systems, but there is currently little mechanistic insight behind the choice of plant cultivars and microbial inoculants. This lack of precision occurs because of two major knowledge gaps: (1) It is unclear what strategies plants use to recruit beneficial microbes, and how much genetic variation exists for this trait; and (2) There is insufficient knowledge of which particular microbes are best partners for boosting plant nutrition from organic sources of N, P, and S.

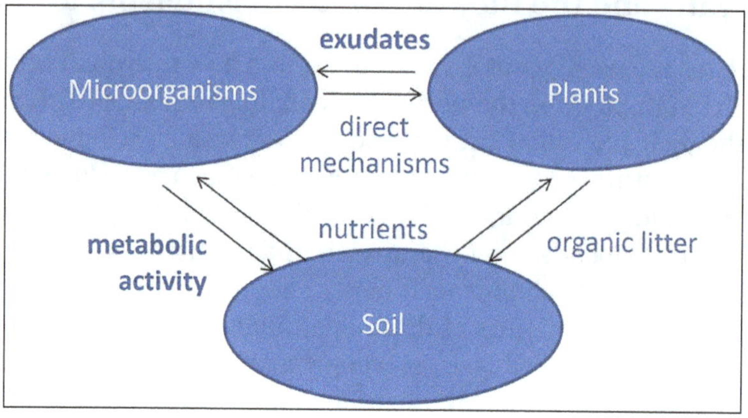

Interactions between plants, microbiota, and soil. Both plants and microorganisms obtain their nutrients from soil and change soil properties by organic litter deposition and metabolic activities, respectively. Microorganisms have a range of direct effects on plants through, e.g., manipulation of hormone signaling and protection against pathogens. Plants communicate with the microorganisms through metabolites exuded by the roots. The major knowledge gaps for understanding the mechanisms of plant–microbe interactions in the rhizosphere are shown in bold.

Effects of Plant Natural Variation on the Rhizosphere Microbiome

To selectively breed plants for optimized nutritional interactions with soil microbes, the genetic components of this trait must first be discovered. Sequence analyses showed differences between the composition of bacterial taxa in soil and plant rhizosphere or the endophytic fraction, showing that plants select for specific bacterial taxa and thus exert some control over their microbiomes. The next question is then to define the key genetic determinants that underpin how different plant genotypes interact with rhizospheric bacteria. Decades of research have shown that the susceptibility to pathogenic microorganisms is highly dependent on plant genome, between different species as well as within accessions of the same one. Similarly, Arabidopsis accessions showed large variation in supporting growth of rhizospheric bacterium Pseudomonas fluorescens in a hydroponic system. Indeed, sequence analyses have confirmed different microbiome structures across plant taxa, with bigger differences in more distant species and with also a larger contribution of environment and soil to the variation. When comparing accessions or varieties of the same species, genotypic effects on microbiome structure have been seen amongst Arabidopsis, maize, and barley. Regarding leaf microbiota, clear differences were shown for leaf microbiomes across 196 Arabidopsis accessions. The variation driven by plant-genome was particularly high for the most abundant operational taxonomic units (OTU). The variation was further explored by genome-wide association study (GWAS) using the number of reads for individual OTUs as quantitative phenotypes. There, GWAS revealed that many of the significant SNPs linked to bacterial OTU structure were categorized as defense response, which was the most overrepresented gene ontology term among the candidate genes. In addition, genes involved in cell wall synthesis and kinase activity were enriched. Although several candidate genes affecting the leaf microbiome were identified, further confirmatory tests of mutants of these genes have not been reported and therefore the functionality of the genes in shaping the microbiome remains to be demonstrated. The leaf microbiome GWAS can be of major importance for understanding the processes in rhizosphere, because the leaf and root microbiomes are overlapping and might be shaped by similar processes. In an alternative approach to GWAS, Bodenhausen et al. monitored changes to SynComs inoculated onto leaves of Arabidopsis accessions and mutants of a priori selected genes. Again, a clear genotype effect upon microbial taxonomic composition has been observed in the 10 accessions and in several mutants. In three mutants, the effects were consistent and reproducible; two mutants were involved in cuticle synthesis and one in ethylene signaling (ein2). Given that only some 40 mutants and highly simplified SynComs were tested, the approach seems to be promising for analysis of root microbiome as well, particularly if mutants in nutrient uptake and assimilation would be investigated.

GWAS of Bacteria-mediated Plant Traits

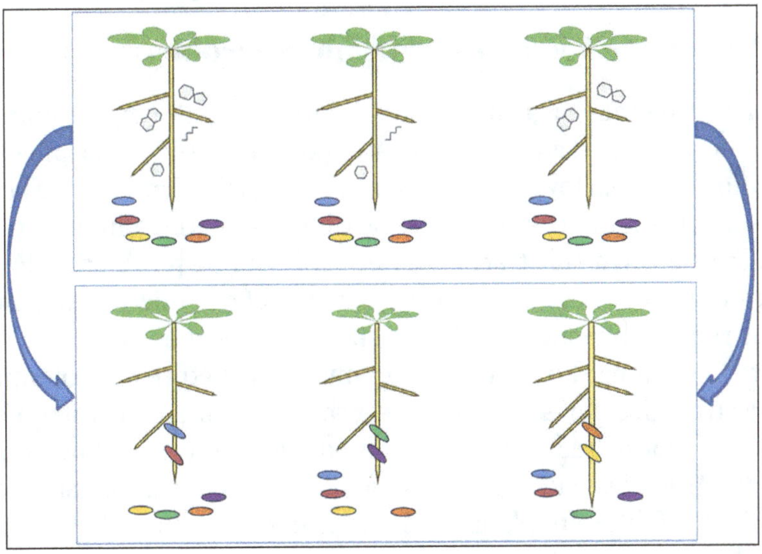

The sequence analyses, however, explore only the taxonomical composition of plant microbiome without taking into account the whole bacterial genomes or addressing the functions these microbes are performing. The best attempt so far to assess how plant genotype affects functional interaction with rhizobacteria is the analysis of variation in susceptibility of Arabidopsis accessions to the plant growth-promoting rhizobacterium Pseudomonas simiae WCS417r. The accessions showed large difference in all three phenotypes scored: fresh weight gain, proliferation of lateral roots, and elongation of the primary root. Statistical GWAS analysis resulted in several highly significant associations, but despite some good correlation between the fresh weight and root architecture data, none of the positive SNPs were found with multiple phenotypes. The analysis led to identification of several candidate genes, but without further verification or confirmatory experiments. The analysis proves that GWAS of Arabidopsis accessions is a feasible approach to identify genetic loci that control the phenotypic variation in plant–microbe interactions. The challenge is to step beyond the relatively simple traits analyzed so far and to design screens that would allow dissecting the genetic architecture of the complex signaling and metabolic networks leading to variation in composition of root associated microbiota in different plant genotypes.

Plant natural variation in root exudates leading to distinctive microbial communities. We postulate that plant genotype has a strong effect on microbial community composition, mediated via root exudate composition. Consider three plant genotypes with differing root exudate profiles. At time 0 (upper panel), the three plants are transplanted from a sterile system onto a common soil where they are proximate to the same set of microbial strains. At time t (lower panel), the three plant genotypes have recruited distinctive microbial strains to their roots, which confer differential growth-promoting effects manifesting in different plant sizes.

Plant Root Exudates—A Source of Molecular Signals

Metabolic Signals to Recruit Favorable Microbes

The growth of soil microbes is usually carbon-limited, so the high amounts of sugars, amino acids, and organic acids that plants deposit into the rhizosphere represent a valuable nutrition source. However, deposition of this labile carbon does not necessarily foster the recruitment of favorable microbes, because pathogenic strains can also use these molecules as growth substrates. Therefore, it can be postulated that plants have evolved recognition mechanisms to discriminate beneficial microorganisms from those that need to be repelled. In such a case, the specific molecules present in root exudates that contribute to shaping the microbial community structure are potential targets for plant breeding strategies that seek to engineer the rhizosphere microbiome. It has been shown that plant root exudates contain components used in belowground chemical communication strategies, such as flavonoids, strigolactones, or terpenoids. Studies on the microbiome of different plant species and accessions revealed strong variations, leading to the hypothesis that exudates are crucial in shaping plant–microbe interactions. Furthermore, it has been shown that plants specifically attract beneficial interaction partners via root derived signals.

Up to now, most information about signal perception and transduction in plant–microbe interactions comes from the field of plant pathology, where plant receptor-like kinases (RLKs) play a major role. In case of mutualistic interactions, nodulation and mycorrhizal interactions serve as model systems to identify recognition mechanisms between plants and microbes. In parallel to recognizing the microbial interaction partner by the plant, also microbes have to recognize their mutual interaction partner (the plant root). It is widely accepted that root exudates contribute to the establishment of the root microbiome. The term "root exudates" describes the molecules that are selectively secreted by roots and distinguishes it from the sloughing-off of root border cells. The overall release of fixed carbon compounds (border cells and exudates) into the surrounding soil is termed as rhizodeposition. Data about the amount of rhizodeposition range between 5 and 30% of the total amount of fixed carbon, which generally means a large loss of reduced-C for biomass and represents a considerable impact on the carbon budget of individual plants and also entire ecosystems. In a 14C approach, Hütsch et al. found remarkable differences in the amount of C-release among six different plant species ranging from 11.6 (wheat) to 27.7 (oil radish) mg C/g root dry matter. Also, the composition of these exudates varied between species, with oil radish exudates being rich in organic acids whereas pea exudates are rich in sugars. These data indicate that various plant species differentially modulate the chemical composition of their rhizospheres, which in turn might impact the associated microbial community. The recruitment of beneficial microbes might be crucial under environmental stress conditions such as nutrient limitation, pathogen attack, pests, high salt, or heavy metal stress.

Issues to Consider when Analyzing Root Exudates

To fully understand the dynamic interactions between soil microbes and plant roots, it is necessary to elucidate the specific molecules within root exudates that can recruit favorable microbial strains. This is a challenging problem in analytical biochemistry, because various biological and methodological issues must be addressed to undertake biologically insightful analyses of plant root exudates. Regarding cultivation, artificial plant growth systems cannot mirror the natural conditions in soil, but on the other hand, it is difficult to unravel the relevant communication signals occurring in soil, due to chemical interaction of metabolites with the soil matrix, and background metabolites released from decomposing organic matter or microbial exudation. Most analyses therefore settle on hydroponic cultivation, sometimes with an inert material to scaffold the roots. When sampling, the experimenter must choose whether to collect exudates in simple deionized water, or a more realistic medium containing mineral salts. Furthermore, it is effectively impossible to design an experimental approach that can differentiate exudates from sloughed-off border cells. A comprehensive summary on exudate collection and influences (e.g., pH, re-uptake by roots, incubation period) is presented in Vranova et al. For data acquisition, researchers are increasingly using unbiased mass spectrometry (MS) approaches such as gas chromatography (GC)-MS and liquid chromatography (LC)-MS. However, detection of all metabolites in a sample is impossible due to physiochemical biases imposed by the selected extraction method, sample clean-up procedure, matrix effects and analytical technique. Therefore, different methods have to be combined for a comprehensive view on the metabolite profile.

The subsequent analysis of the derived MS data is a huge challenge, beginning with data processing algorithms that enable feature detection, peak alignment and different normalization methods. These normalization and scaling algorithms have a large impact on the outcome of an analysis. To validate the identity of specific mass spectral features, fragmentation data (MS^2 or MS^n) are acquired and compared against publicly available databases, or authentic standards. Taken together, these challenges mean that comprehensive analysis of root exudates is not trivial.

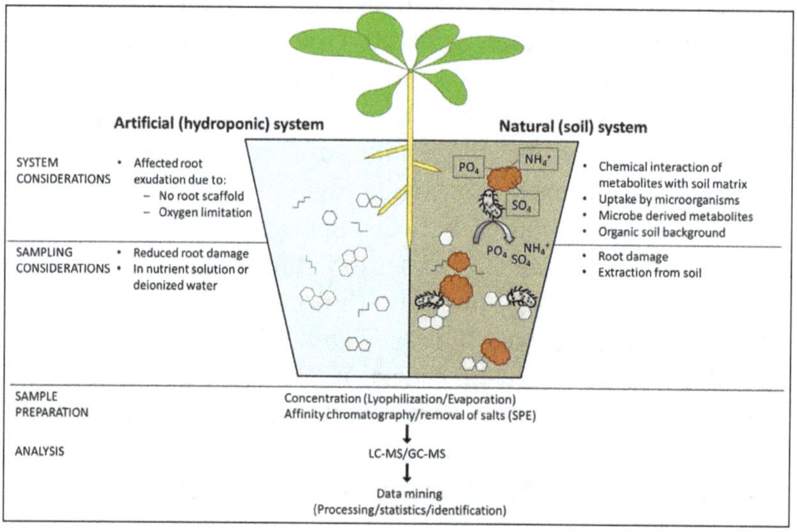

Technical considerations for analyses of root exudates. Comprehensive analyses of root exudate composition are crucial to advance our knowledge of plant–microbe interactions, but experimenters must consider the technical challenges at each step of the analytical workflow. First the growth system must be carefully considered, particularly whether to use soil or hydroponics, which each have advantages and disadvantages. At sampling, those using hydroponic systems must decide whether to collect the exudates in nutrient solution or deionized water, whereas those using soil systems must consider how to separate exudates from the soil matrix. During sample preparation, there are a multitude of options for sample concentration and clean-up, which will influence sample composition. Clearly, the mass spectrometry methodology used for data acquisition will play a crucial role in the workflow, because different setups allow the detection of different molecules. Finally, there is a growing awareness that data processing and analysis strategies also play a key role in shaping the derived data.

Recent Approaches to Analyze Root Exudate Composition

Various studies have described analyses of plant root exudates, with Phillips et al. developing a method to collect exudates from mature trees in the field, although microbial metabolism probably makes a significant impact upon this non-sterile system. That said, microbial nutrient uptake is an interesting aspect of plant–microbe nutritional interactions, with a fast degradation of flavonoid glucosides being observed by Carlsen et al. when comparing the flavonoid content in two soils and after different legume cultivations. To avoid microbial impact upon root exudate profiles, researchers have established diverse approaches of axenic hydroponic cultivation systems, which are easier to control, even though they represent artificial plant cultivation systems and plant responses might also include stress reactions due to oxygen limitation and insufficient root support.

Furthermore, hydroponics is well suited for sampling of exudates, as the total liquid can directly been taken for further sample preparation procedures and root damage is minimized. However, collection of exudates widely ranges in timescale and the used collection medium (nutrient solution or water). Badri et al. collected root exudates from Arabidopsis thaliana in nutrient solution for 3 and 7 days for analysis by LC-MS and revealed that most compounds are present only after the longer incubation period. It could be hypothesized that this observation is due to sloughed-off border cells. Nevertheless, they also compared the exudate composition against root composition and stated an 80% difference based on detected molecular masses.

Also Strehmel et al. applied a 7-day collection period in nutrient solution to obtain sufficient amounts of exudates from A. thaliana. In contrast, Carvalhais et al. applied only a 6-h exudate collection period to Zea mays plants to minimize the effect of sloughed-off border cells, but used deionized water as collection medium. A similar approach has been used for barley root exudates that were collected for 4 h in deionized water. For a short-term exudate collection period from Arabidopsis a high amount of plants has been required to obtain sufficient amounts of exudates for LC-MS analysis. A direct comparison of different plant cultivation and exudate collection techniques revealed a huge impact on metabolite patterns.

Especially, long incubations in deionized water may lead to overestimated exudation rates due to the high transmembrane gradient of solutes in low ionic strength medium. To date, most published data on exudates concentrate on specific metabolite classes such as primary metabolites, hormones, flavonoids, or phytosiderophores. Non-targeted metabolite profiling approaches of root exudates have been applied less frequently, although Strehmel et al. provided a comprehensive overview on secondary metabolites in Arabidopsis root exudates using LC-MS. In follow-up experiments, the data collection was complemented by GC-MS data and extended by a comparison of 19 Arabidopsis accessions, co-cultivation with Piriformospora indica and data from phosphate limitation. As MS technology continues to improve, it can be expected that more studies will undertake untargeted analyses of root exudate profiles.

How Root Exudates Differ across Plant Genotype and Nutrient Limitation

If root exudate profiles are to be a potential breeding target for increasing plant–microbe nutritional cooperation, then it must be first understood how exudate composition varies across genotypes or in response to nutrient deprivation. Recent studies have contributed to our knowledge of this phenomenon, with the effects of phosphate limitation investigated in Ziegler et al., and variation across accessions shown in Monchgesang et al. and also Micallef et al. Comparing exudate profiles across 19 natural Arabidopsis accessions showed a high natural variation for glycosylated and sulfated metabolites, such as flavonoids, glucosinolate degradation products, salicylic acid catabolites and polyamine derivatives. Regarding root exudation changes induced by nutrient limitation, it seems that phosphate deficiency results in a higher abundance of oligolignols and a lower abundance of coumarins. Future experiments could investigate a different panel of plant genotypes, perhaps where previous experiments have defined a phenotypic difference that is potentially linked to root exudation profiles. One possibility involves comparing genotypes that have shown contrasting affinity to recruit favorable microbes, or genotypes that differ in their nutrient starvation responses. Another option is root exudate profiling to analyze the phenotypic effects of mutants that were identified by GWAS studies.

Specific Molecules in Root Exudates Linked to Plant–microbe Nutritional Interactions

From our current knowledge of root exudates, is it possible to pinpoint a set of molecules that are particularly promising for recruiting favorable microbes to the rhizosphere? In legumes, it is well described that the flavonoid pathway has a huge impact on attracting rhizobia bacteria to roots and inducing NOD gene expression. Flavonoids are also crucial for hyphal branching and thus promoting mycorrhizal interaction. Both of these interactions result in increased plant nutrient uptake, with mycorrhiza and root nodules boosting phosphorus and nitrogen, respectively. Perhaps other plants and soil bacteria have also tapped into this signaling pathway, and further analysis of root exudates will give us clues as to whether flavonoids play a role in this communication leading to increased nutrient uptake. From analyzing plant mutants, candidate genes that have been investigated so far are related to the transfer of metabolites into the rhizosphere, hormonal signaling, or to biosynthesis, e.g., genes from phenylpropanoid pathway. These results give us some hints about candidate molecules that are exuded by plant roots to recruit beneficial bacteria, such as strigolactones and flavonols, and presumably further analyses will expand this list. To fully exploit this knowledge, then it is desirable to identify which microbial strains are recruited by these molecules, and what benefits they confer to the plant.

Using Sequence Data to Predict Microbial Effects on Plant N, P and S Nutrition

Mechanisms of Microbial Nutrient Provision to Plants

Plants give large amount of carbon away to the rhizosphere that nourishes soil microorganisms. So what do plants get back? In natural soils the vast majority of

N, P, and S atoms are organically bound, while in the atmosphere the vast majority of N is contained in the N_2 molecule. Due to the different metabolic capacities of plants and microbes, these nutrient sources are minimally bioavailable to plants, but can be metabolized by various soil microbes. This means that nitrogen fixing and nutrient mineralization processes carried out by soil microbes are crucial for plant nutrition in natural ecosystems, because these reactions metabolize recal-citrant forms of N, P, and S to liberate these elements for plant nutrition. It must be briefly stated that this established paradigm has been somewhat questioned in recent years, as several studies have demonstrated direct plant uptake of various organic-N forms. However, it is still generally accepted that microbes are better competitors for these nutrients due to the low diffusivity of organic-N molecules in soil, and the results of isotope labeling studies generally support the concept that most organic-N is first assimilated by microbial taxa, then subsequently assimilated by plants upon microbial turnover.

Over several decades, the soil microbiology literature has accumulated a list of microbial metabolic processes that are linked to plant N, P, and S nutrition. Commercially, the symbiotic association between legumes and bacteria is routinely exploited when field crops are inoculated with nitrogen-fixing rhizobia strains. But, how can this phenomenon be refined and optimized for widespread use in more sustainable agricultural systems—not only for nitrogen fixation in legumes, but also for N, P, and S nutrition in non-leguminous crops? Given that different bacterial strains exhibit differing metabolic capabilities, coupled with the huge amount of genomic sequence data from soil microbes that has recently been generated, one possibility is to pinpoint the genes that encode metabolic pathways for agriculturally beneficial N, P, and S metabolism, and to boost the microbes that contain these specific genes in agricultural soils.

Table: Key microbial metabolic processes related to plant nutrition.

Element	Biochemical Process	Microbial Genes
Nitrogen	Nitogen Fixation Protein Depolymerization Urea Catabolism.	nifD, nifH, nifK apr, npr, sub ureA, ureB, ureC.
Phosphorous	Phosphate Ester Cleavage Phosphonate Breakdown.	phoA, phoD, phoX, ACPase, glpQ, ushA, appA, phyA, phyB phnJ, phnX.
Sulphur	Sulphate Ester Cleavage Sulphonate Breakdown.	aslA, asfA ssuD.

Cultivation-independent or Cultivation-dependent Approaches

The major difficulty in investigating soil microbial communities is that only a small fraction of the inhabiting taxa can be cultivated in the laboratory, where experimenters can undertake detailed and controlled analyses. Generally, most cultivation-independent approaches extract root-associated microbes in situ, and then analyze the properties of this community. In contrast, cultivation-dependent approaches generally inoculate soil or root-associated microbes onto laboratory growth medium, before analyzing distinct strains that have been cultivated in the laboratory. For cultivation-independent approaches, techniques typically used to analyze soil microbiota often include: (1) 16S sequencing and PLFA measurements to infer taxonomic breakdown; (2) Metagenomic, metatranscriptomic, or metaproteomic analyses to infer functional capacity of microbial communities; and (3) Enzyme assays, respiratory measurements, or substrate utilization assays to measure functional activity of microbial communities. In cultivation-dependent approaches, the analytical possibilities are multitude—once root-associated organisms have been cultivated in a laboratory setting, they can be analyzed with any available technique. There are many studies that use both cultivation-dependent and cultivation-independent techniques to draw links between microbial properties and plant nutrition. But, can these acquired data inform the rational selection of microbes that will improve plant N, P, and S nutrition?

Microbial Genes in Cultivation-dependent Literature Linked to Soil Fertility

It can be posited that soils with high rates of microbial N, P, and S cycling should harbor microbes with specialized genes encoding these traits, and therefore that microbial strains isolated from these soils should possess useful metabolic attributes for boosting plant nutrition. Interestingly, Rhizobia strains isolated from N-fertilized soils exhibited a lower capacity to promote plant growth compared to strains isolated from adjacent unfertilized plots, indicating that the management history of the isolation site impacts the degree of mutualism in the resulting isolate. In the P literature, a similar phenomenon has been observed, with phosphate mineralization being more common in isolates from soils where bioavailable-P was less abundant. For sulfur, one example is from soils from the Rothamsted Broadbalk experiment, where different S-fertilization practices had led to fields that exhibit high versus low sulfatase activity. Bacterial strains were isolated from these contrasting soils, and functional assays such as enzyme measurements and growth on minimal media revealed that the strains isolated from low-SO_4^{2-} soils contained several mechanisms for depolymerizing organic-S. Together, these results imply that research programs seeking "elite" microbial strains that can maximally boost plant nutrition could begin with inocula from sites that favor plant–soil feedbacks, such as unfertilized soils or organic farms.

Microbial Strains that Promote Plant Growth by Enhancing N, P, and S Nutrition

To truly be useful in an agricultural setting, it must be proven that candidate growth-promoting strains can be re-inoculated onto plants, successfully colonize the rhizospheric niche, and then mediate nutrient mobilization that benefits plant growth.

This can be tested through plant–microbe interaction assays, where candidate strains are tested for their ability to promote plant growth and nutrient acquisition. Once again, this research field is most mature for the case of nitrogen-fixing Rhizobia, where decades of research have endeavored to define the optimal inoculation practice, searching for the right combination of plant genotypes and rhizobia strains to suit specific climates and soils. Regarding the taxonomy of nitrogen-fixing symbioses, it should be mentioned that nitrogenase genes are present in diverse bacterial taxa and that non-leguminous plants have been documented to host N_2-fixing bacterial strains, perhaps implying that other plant–microbe combinations (not just legumes and Rhizobia) could be similarly optimized to promote nitrogen fixation. There are also reports of plant growth promotion via microbial mobilization of other nitrogen sources, shown by higher yield in plants inoculated with bacterial strains, although for one of these experimental setups, it seems that the source of this N was directly from the ammonium sulfate fertilizer rather than from organically bound soil N. It has been shown that unsterilized grass seeds can better access protein-N compared to sterilized seeds, but the specific strains that provide this service were not elucidated.

The ability of the fungi Glomus intraradices to transfer organic nitrogen to plants has also been shown, suggesting that future experiments could focus on documenting other fungal strains with this capacity and characterizing the relevant genes and mechanisms. For phosphorous, the literature contains a large number of reports of both fungal and bacterial strains with the capacity to solubilize inorganic P, and also many reports of strains that can mineralize organic P. Many of these P-mobilizing strains are also characterized as growth promoting microbes, but microbial promotion of plant growth can operate through a wide variety of mechanisms, and sometimes it is not conclusive that P-mobilization is responsible for the plant growth promotion elicited by these strains. For sulfur, studies of a plant growth-promoting Pseudomonas strain have used genetic knockout of the sulfonate monooxygenase enzyme to show that organic-S mineralization accounts for some fraction of the growth-promoting phenotype.

The research field is beginning to build large collections of genomically sequenced bacterial isolates that can be re-assembled into SynComs isolated endophytic bacterial strains from plants grown under organic management, and showed that over half of these strains can boost tomato growth in a greenhouse experiment. This high proportion of growth-promoting isolates shows that the capacity to promote plant growth is widespread amongst plant-associated bacteria, but to gain a mechanistic understanding into how these growth-promoting effects are manifested, it will be necessary to conduct detailed investigations into the genetics, biochemistry and physiology of these growth-promoting strains. Furthermore, microbial community experiments should also consider how the interaction between different strains affects plant growth promotion. Such knowledge will enable the rational selection of growth-promoting strains and communities, driven by defined genetic and biochemical mechanisms.

Growing Community: Challenges and Perspectives in Modeling Plant–microbe Interactions

With the vast quantity of data being generated by high-throughput experimental techniques, new opportunities are arising to integrate theoretical and computational approaches with experiments. Potential synergies include designing hypothesis-driven experiments based on the results of modeled scenarios, or using modeling as a tool to mechanistically interpret the results of high-throughput experiments. In the bacterial field, there is currently a major initiative to integrate genomic sequencing data with computational modeling, in order to predict the function of individual bacterial strains and whole bacterial communities. Already, microbial engineering is a mature technology that exploits the power of computational modeling to optimize strains and communities for industrial processes like bioremediation and fermentation. Regarding the study of the rhizosphere microbiome, over the past 15 years this field has accumulated a huge volume of sequencing data, but these data are generally descriptive, and to date they have not been used to significantly further our mechanistic understanding of nutrient exchange processes in the rhizosphere.

One of the principal aims of mathematical models is the reduction of complexity in order to capture the fundamental principles behind the phenomena of interest. There is nowadays a positive trend in biology to develop predictive models of the system under study, with some disciplines at a more advanced stage than others in the integration of theory and experiments. The study of microbial communities is a clear illustrative example of a field where computational approaches are essential. While bioinformatics is instrumental to analyze high throughput data from meta-omics experiments, theoretical models are developed to gain mechanistic understanding of complex biological systems. Mathematical descriptions of population dynamics were pioneered in the 19th century by Verhulst's law of logistic growth and are now a fundamental part of ecology. In the same way, over the last five decades models of metabolism have started converging into sound mathematical methods to investigate cellular functions. It is also relevant to point out that another characteristic of mathematical models is that, as long as the described mechanism holds true, they can be easily generalized to different organisms. Therefore, as an example, a model originally built to describe bacteria might be also applied to other microbes.

Theoretical biology is a rapidly expanding field and it is nearly impossible to condense and classify it in few words, but we can roughly point out three main classes of methods: kinetic models, stochastic models, and network-based models. It is important to point out that depending on the system and phenomena under study, certain modeling techniques will be more suited than others. Kinetic models are dynamic and deterministic, typically constructed as systems of differential equations, solved nowadays with computational integration algorithms. These models can offer precise predictions but require either a priori knowledge or inference (e.g., through a fit to data) of the equation parameters. Typically, these models are well suited to describe small-scale metabolic pathways where most enzyme kinetic constants are measured and few parameters are reasonably constrained and then fit. Stochastic models encompass any representation that implements some random components with a Monte Carlo procedure and they are needed to capture effects like noise or individual variability. A widely used method is the Gillespie algorithm which simulates in randomized steps the evolution in time of, e.g., a chain of biochemical reactions, each of which has an associated probability to happen. Under the general term network-based models we include different static methods that have in common the treatment of metabolic pathways as networks, where metabolites are connected to reactions either as substrates or products. Reactions can be represented either as edges connecting nodes (the metabolites) or, in bipartite networks, as a disjoint set of nodes (the enzymes) that connect to the metabolites nodes via edges carrying, e.g., stoichiometric information. Reversibility of a reaction is determined by physical principles and is taken into account in directed networks.

Mostly, systems of plant–microbes have been considered in light of host–pathogen interactions. The classic zig-zag model is an illustrative scheme proposed to explain the function of the plant immune system in response to pathogens. It distinguishes two branches of the plant immune system, one reacting to microbe-associated molecular patterns with pattern-triggered immunity (PTI), the other responding to effectors trying to suppress PTI with effector-triggered immunity. This picture is, however, a purely expository model and is not suited to capture the interaction dynamics we are interested in. Indeed, if the objective is to build a predictive model of plant–microbe interactions, and in particular beneficial ones, the fundamental ingredients include: (i) The actual molecular factors driving the interactions; (ii) Environmental conditions inducing different reactions, e.g., to stresses; (iii) Temporal and spatial scales of the phenomena under study. While it would be nice to have a single model drawn from a universal theory of plant–microbe interactions, we are mostly at the stage where mathematical models are first needed to answer a limited number of focused and well-defined biological questions. With the current advancement in "omics" experimental techniques and with the development of computational methods to understand metabolic pathways, quantitative models of plant–microbes ecosystems at the molecular level become an appealing possibility.

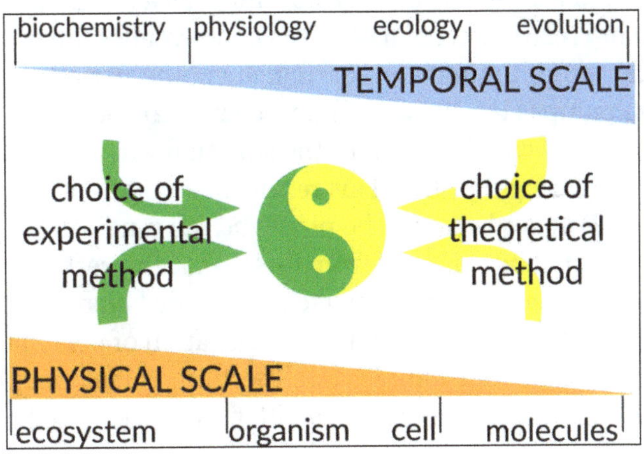

Considerations for combining modeling and experimental approaches. A major goal in biology is to integrate computational predictions with experimental data to generate predictive models of biological systems. While the temporal scale is an intrinsic property of the phenomena under study, the choice of the physical scale of the model/experiment is chosen by the scientist: one can investigate soil microbial communities at an ecosystem macroscale or at the DNA sequencing level. In general, the physical scale will have a strong impact on the choice of experimental technique, while the temporal scale will mostly influence the experimental design (this is represented by thicker/thinner arrows). On the contrary, theoretical methods will be highly constrained by the temporal scale to be modeled, while they are rather flexible regarding the physical scale to be operated at. In a truly interdisciplinary approach, the experimental and theoretical methods have to be planned together to ensure that the reciprocal results are compatible and can be integrated. This will allow further improvements in both experimental design and model development.

Dynamic Models of the Rhizosphere

The rhizosphere comprises the shared environment between the plant roots and microbes. Understanding the dynamics of nutrient competition in the rhizosphere is an important aspect to identify the driving mechanisms of community assembly. The work of Darrah presents one of the first differential equation models of microbial population dynamics in the rhizosphere, parameterizing how root growth and exudation of soluble C fuels microbial growth at specific sites. These models accounted for root elongation and release of soluble organic compounds which diffuse through the surrounding area populated by bacteria. The simulations showed how different exudation patterns can control the distribution of bacterial biomass. Being based on simple equations representing the spatial gradient of soluble substrates, this model is easily expanded to include the presence in the rhizosphere of other solutes like minerals essential to plant growth. Recently, a similar approach was used to model the colonization of the root tip by bacteria, introducing bacterial motility and under the assumption that carbon is the growth-limiting nutrient.

Borrowing concepts from microeconomics, Schott et al. modeled the nutrient trade system between plants and arbuscular mycorrhizal fungi as a minimal network, and investigated the partnership relation in terms of costs and revenues. First, they demonstrated that a simple model where the nutrient exchange via the periarbuscular space is only regulated by proton pumps (H^+-ATPase) and nutrient transporters (H^+/sugars and H^+/phosphate cotransporters) could reproduce the situation where C and P exchange is stable at the thermodynamic equilibrium. This result shows that this small set of transporters can explain C and P fluxes between the plant and the fungi, although the presence of further transporters cannot be excluded from this result. Interestingly, while the observable highly cooperating behavior might suggest altruistic behavior in a "reward-based" system, the symbiosis is the result of a selfish optimization of personal benefit. A change in the costs to revenues ratio, for example, a well-fertilized plant that places a lower value on fungal-derived P would therefore lead to a different nutrient exchange rate.

Soil nutrient fluxes also play a role in more global scenarios. Earth system models are extended climate models that include the effects from biogeochemical cycles, and different competition theories can be embedded in them. Recently, Zhu et al. demonstrated that a kinetic model considering plant, microbes and nutrients as a coupled network, and based on the principle that nutrient uptake rate is controlled by specialized transporter enzymes, can reproduce data of N competition in a grassland ecosystem where other competition theories fail.

Genome-scale Metabolic Network Models

Differential equation models are perfectly suited to simulate dynamic processes like enzymatic or diffusion reactions, but rely on fixed parameters that characterize specific states. In principle it would be possible to obtain exact kinetic models of metabolic networks at the "genome scale" (genome-scale models, GEMs) if we were to know the function and the kinetic parameters of each enzyme in the picture. Such models would be, however, computationally expensive to solve and difficult to manage in terms of number of equations, variables and parameters. Instead, a convenient approach is the reconstruction of GEMs from automated genome annotation and their representation as networks. In an automated process of mining the information from databases, the enzymes encoded in the genome are assigned metabolic functions. The resulting GEM, however, still requires subsequent careful manual curation as a consequence of possible inaccurate annotations besides our incomplete knowledge of genes associated to metabolic activities.

Once a metabolic network has been reconstructed, it can be translated into a stoichiometric matrix. This mathematical formalism, connecting reactions to metabolites via the stoichiometric coefficients, presents itself to a number of methods developed to study the function of biochemical pathways. Constraint-based models investigate the distribution of reaction fluxes along the directed metabolic network (first constraint, imposing irreversibility or reversibility under physiological conditions from thermodynamics principles) under the assumption that the system is at a steady state (second constraint, meaning that there is internal balance of production and consumption of metabolites) and that enzymes operate at limited capacities (third constraint, setting upper or lower bounds to the rate of an enzymatic reaction based on experimental observations). This set of constraints leads to a narrower space of possible solutions for the metabolic fluxes, but a further step is needed in order to identify a unique solution. A widely used method is flux balance analysis (FBA), where an optimization problem is defined by introducing an objective function (usually a linear combination of metabolic rates) either to be maximized (like in the case of biomass as objective function) or minimized (like in the case of total sum of cellular fluxes).

GEMs are in general compartmentalized according to the cellular structure. The simplest division, typically for prokaryotic organisms, is between an external and an internal (cytosolic) compartment, while plant GEMs will have compartments for different tissues as well. Exchange of metabolites between compartments happens through transport reactions, which are not always correctly annotated and often added to the network during the gap-filling process of the reconstruction. An exchange flux with the external compartments is equivalent to an uptake or secretion rate.

The GEM of the nitrogen fixing bacterium Rhizobium etli, a known symbiont of legumes, was studied with constraint-based modeling to understand its physiological capabilities during the third developmental stage of the symbiosis in the plant nodule, when it differentiate into a bacteroid and provides N to the plant. For a given oxygen uptake rate, an increase in succinate uptake rate leads to higher symbiotic N fixation, until a threshold value in succinate uptake rate is reached. After that an inhibitory effect, caused by insufficient oxygen for succinate reduction, drastically lowers N fixation rate. These kinds of studies, quick to perform in silico, can be easily transferred to other pathways of interest to generate hypothesis relating environment to metabolism.

Available GEMs can be browsed at the BioModels database, where most of the reconstructed models are deposited and usually classified according to their level of curation. We have to point out that efforts are still on-going to converge on model standards, and the trustworthiness of simulation results depends on the quality of the GEM.

Soil Microbes and Nutrient Recycling

Soil microorganisms exist in large numbers in the soil as long as there is a carbon source for energy. A large number of bacteria in the soil exists, but because of their small size, they have a smaller biomass. Actinomycetes are a factor of 10 times smaller in number but are larger in size so they are similar in biomass to bacteria. Fungus population numbers are smaller but they dominate the soil biomass when the soil is not disturbed. Bacteria, actinomycetes, and protozoa are hardy and can tolerate more soil disturbance than fungal populations so they dominate in tilled soils while fungal and nematode populations tend to dominate in untilled or no-till soils.

There are more microbes in a teaspoon of soil than there are people on the earth. Soils contain about 8 to 15 tons of bacteria, fungi, protozoa, nematodes, earthworms, and arthropods.

Table: Relative number and biomass of microbial species at 0–6 inches (0–15 cm) depth of soil.

Microorganisms	Number/g of soil	Biomass (g/m^2)
Bacteria	10^8–10^9	40–500
Actinomycetes	10^7–10^8	40–500
Fungi	10^5–10^6	100–1500

Algae	10^4–10^5	1–50
Protozoa	10^3–10^4	Varies
Nematodes	10^2–10^3	Varies

Microbial Soil Organic Matter Decomposition

Organic matter decomposition serves two functions for the microorganisms, providing energy for growth and suppling carbon for the formation of new cells. Soil organic matter (SOM) is composed of the "living" (microorganisms), the "dead" (fresh residues), and the "very dead" (humus) fractions. The "very dead" or humus is the long-term SOM fraction that is thousands of years old and is resistant to decomposition. Soil organic matter has two components called the active (35 percent) and the passive (65 percent) SOM. Active SOM is composed of the "living" and "dead" fresh plant or animal material which is food for microbes and is composed of easily digested sugars and proteins. The passive SOM is resistant to decomposition by microbes and is higher in lignin.

Microbes need regular supplies of active SOM in the soil to survive in the soil. Long-term no-tilled soils have significantly greater levels of microbes, more active carbon, more SOM, and more stored carbon than conventional tilled soils. A majority of the microbes in the soil exist under starvation conditions and thus they tend to be in a dormant state, especially in tilled soils.

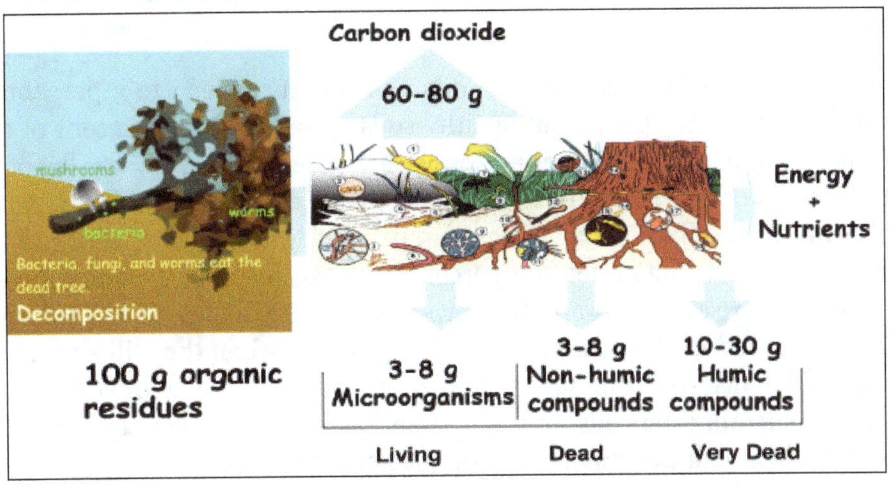

Dead plant residues and plant nutrients become food for the microbes in the soil. Soil organic matter (SOM) is basically all the organic substances (anything with carbon) in the soil, both living and dead. SOM includes plants, blue green algae, microorganisms (bacteria, fungi, protozoa, nematodes, beetles, springtails, etc.) and the fresh and decomposing organic matter from plants, animals, and microorganisms.

Soil organic matter can be broken down into its component parts. One hundred grams (g) or 100 pounds (lbs) of dead plant material yields about 60–80 g (lbs) of carbon dioxide, which is released into the atmosphere. The remaining 20–40 g (lbs) of energy and nutrients is decomposed and turned into about 3–8 g (lbs) of microorganisms (the living), 3–8 g (lbs) of non-humic compounds (the dead), and 10–30 g (lbs) of humus (the very dead matter, resistant to decomposition). The molecular structure of SOM is mainly carbon and oxygen with some hydrogen and nitrogen and small amounts of phosphorus and sulfur. Soil organic matter is a by-product of the carbon and nitrogen cycles.

Soil Organic Matter Nutrients

The nutrients in the soil have a current value of $680 for each 1 percent SOM or $68 per ton of SOM based on economic values for commercial fertilizer. SOM is composed of mostly carbon but associated with the carbon is high amounts of nitrogen and sulfur from proteins, phosphorus, and potassium. SOM should be considered like an investment in a certificate of deposit (CD). Soils that are biologically active and have higher amounts of active carbon recycle and release more nutrients for plant growth than soils that are biologically inactive and contain less active organic matter. Under no-till conditions, small amounts of nutrients are released annually (like interest on a CD) to provide nutrients slowly and efficiently to plant roots. However, with tillage, large amounts of nutrients can be released since the SOM is consumed and destroyed by the microbes. Since SOM levels are slow to build, the storage capacity for nutrients is decreased and excess nutrients released are often leached to surface waters. SOM is a storehouse for many plant nutrients.

Consider the following three scenarios. Soils typically turnover 1 to 3 percent of their nitrogen stored in SOM. Tilled or unhealthy soils release a lower percent of nitrogen due to lower microbial activity. A tilled soil with 2 percent SOM (2,000 lbs of N) may release 1 percent N or 20 lbs of N per year. A soil that is more biologically active and has 4 percent SOM (4,000 lbs N) may release 1.5 percent N or 60 lbs N while a 6 percent SOM soil (6,000 lbs N) may release 2 percent N or 120 lbs of N. In tilled soils, excess nutrients released are often lost and the carbon stores are depleted so that future storage of nutrients is reduced. Farmers often see this occur when they till a virgin soil, an old pasture, or a fence row. For several years, crops on the newly tilled soil will grow better than the surrounding soils, but over time the soil will be depleted of carbon and the newly tilled soil will become less fertile because the carbon is oxidized as carbon dioxide and lost to the atmosphere. Tillage results in the oxidation and destruction of carbon in the soil by increasing the soil oxygen levels, thereby promoting bacteria populations to expand and consume active carbon in the soil.

Table: Value of soil organic matter.

Assumptions: 2,000,000 pounds soil in top 6 inches	
Nutrients	1% organic matter = 20,000# 50%Carbon, C:N ratio = 10:1
Nitrogen:	1000# * $0.50/#N = $500
Phosphorus:	100# * $.70/#P = $70

Potassium:	100# * $0.40/#K = $40
Sulfur:	100# * $0.50/#S = $50
Carbon:	10,000# or 5 ton * $4/Ton = $20
Value of 1% SOM Nutrients/Acre	= $680
Relative Ratio of Nutrients:	100 Carbon/10 Nitrogen/ 1 Phosphorus/1 Potassium/1 Sulfur

Climate, Temperature and pH Effects on SOM

SOM is affected by climate and temperature. Microbial populations double with every 10 degree Fahrenheit change in temperature. If we compare the tropics to colder arctic regions, we find most of the carbon is tied up in trees and vegetation above ground. In the tropics, the topsoil has very little SOM because high temperatures and moisture quickly decompose SOM. Moving north or south from the equator, SOM increases in the soil. The tundra near the Arctic Circle has a large amount of SOM because of cold temperatures. Freezing temperatures change the soil so that more SOM is decomposed then in soils not subject to freezing.

Moisture, pH, soil depth, and particle size affect SOM decomposition. Hot, humid regions store less organic carbon in the soil than dry, cold regions due to increased microbial decomposition. The rate of SOM decomposition increases when the soil is exposed to cycles of drying and wetting compared to soils that are continuously wet or dry. Other factors being equal, soils that are neutral to slightly alkaline in pH decompose SOM quicker than acid soils; therefore, liming the soil enhances SOM decomposition and carbon dioxide evolution. Decomposition is also greatest near the soil surface where the highest concentration of plant residues occur. At greater depths there is less SOM decomposition, which parallels a drop in organic carbon levels due to less plant residues. Small particle sizes are more readily degraded by soil microbes than large particles because the overall surface area is larger with small particles so that the microbes can attack the residue.

A difference in soil formation also occurs traveling east to west across the United States. In the east, hardwood forests dominated and tree tap roots were high in lignin, and deciduous trees left large amounts of leaf litter on the soil surface. Hardwood tree roots do not turn over quickly so organic matter levels in the subsoil are fairly low. In forest soils, most of the SOM is distributed in the top few inches. As you move west, tall grassland prairies dominated the landscape and topsoil formed from deep fibrous grass root systems. Fifty percent of a grass root dies and is replaced every year and grass roots are high in sugars and protein (higher active organic matter) and lower in lignin. So soils that formed under tall grass prairies are high in SOM throughout the soil profile. These prime soils are highly productive because they have higher percentage of SOM (especially active carbon), hold more nutrients, contain more microbes, and have better soil structure due to larger fungal populations.

Carbon to Nitrogen Ratio

The breakdown of organic residues by microbes is dependent upon the carbon to nitrogen C:N ratio. Microbes in a cow's rumen, a compost pile, and soil microbes rely on the C:N ratio to break down organic (carbon-based) residues. Consider two separate feed sources, a young tender alfalfa plant and oat or wheat straw. A young alfalfa plant has more crude protein, amino acids, and sugars in the stalk so it is easily digested by microbes whether it is in a cow's rumen, a compost pile, or in the soil. Young alfalfa has a high nitrogen content from protein (amino acids and proteins are high in nitrogen and sulfur), so it has a lower carbon to nitrogen ratio (less carbon, more nitrogen). However, oat and wheat straw (or older mature hay) has more lignin (which is resistant to microbial decomposition), lower crude protein, and less sugars in the stalk and a higher C:N ratio. Straw is decomposed by microbes but it takes additional time and nitrogen to break down this high carbon source.

A low nitrogen content or a wide C:N ratio is associated with slow SOM decay. Immature or young plants have a higher nitrogen content, lower C:N ratios and faster SOM decay. For good composting, a C:N ratio less than 20 allows the organic materials to decompose quickly (4 to 8 weeks) while a C:N ratio greater than 20 requires additional N and slows down decomposition. So if we add a high C based material with low N content to the soil, the microbes will tie up soil nitrogen. Eventually, the soil N is released but in the short-term the N is tied up. The conversion factor for converting N to crude protein is 16.7, which relates back to why it is so important to have a C:N ratio of less than 20.

The C:N ratio of most soils is around 10:1 indicating that N is available to the plant. The C:N ratio of most plant residues tends to decrease with time as the SOM decays. This results from the gaseous loss of carbon dioxide. Therefore, the percentage of nitrogen in the residual SOM rises as decomposition progresses. The 10:1 C:N ratio of most soils reflects an equilibrium value associated with most soil microbes (Bacteria 3:1 to 10:1, Fungus 10:1 C:N ratio).

Bacteria are the first microbes to digest new organic plant and animal residues in the soil. Bacteria typically can reproduce in 30 minutes and have high N content in their cells (3 to 10 carbon atoms to 1 nitrogen atom or 10 to 30 percent nitrogen). Under the right conditions of heat, moisture, and a food source, they can reproduce very quickly. Bacteria are generally less efficient at converting organic carbon to new cells. Aerobic bacteria assimilate about 5 to 10 percent of the carbon while anaerobic bacteria only assimilate 2 to 5 percent, leaving behind many waste carbon compounds and inefficiently using energy stored in the SOM.

Fungus generally release less carbon dioxide into the atmosphere and are more efficient at converting carbon to form new cells. The fungus generally captures more energy from the SOM as they decompose it, assimilating 40 to 55 percent of the carbon.

Most fungi consume organic matter higher in cellulose and lignin, which is slower and tougher to decompose. The lignin content of most plant residues may be of greater importance in predicting decomposition velocity than the C:N ratio.

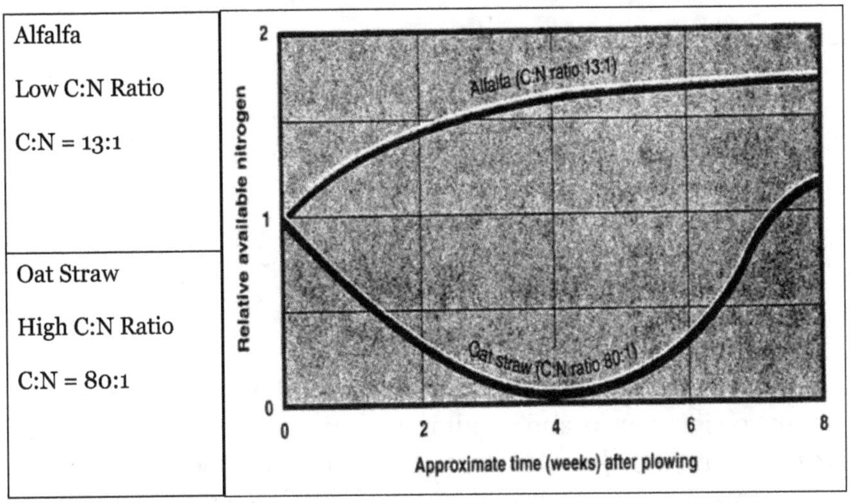

Graph of relative available n with length of time for decomposition.

Mycorrhizal fungi live in the soil on the surface of or within plant roots. The fungi have a large surface area and help in the transport of mineral nutrients and water to the plants. The fungus life cycle is more complex and longer than bacteria. Fungi are not as hardy as bacteria, requiring a more constant source of food. Fungi population levels tend to decline with conventional tillage. Fungi have a higher carbon to nitrogen ratio (10:1 carbon to nitrogen or 10 percent nitrogen) but are more efficient at converting carbon to soil organic matter. With high C:N organic residues, bacteria and fungus take nitrogen out of the soil.

Protozoa and nematodes consume other microbes. Protozoa can reproduce in six to eight hours while nematodes take from three days to three years with an average of 30 days to reproduce. After the protozoa and nematodes consume the bacteria or other microbes (which are high in nitrogen), they release nitrogen in the form of ammonium. Ammonium (NH_4^+) and nitrates (NO_3^-) are easily converted back and forth in the soil. Plants absorb ammonium and soil nitrates for food with the help of the fungi mycorrhizal network.

Microorganism populations change rapidly in the soil as SOM products are added, consumed, and recycled. The amount, the type, and availability of the organic matter will determine the microbial population and how it evolves. Each individual organism (bacteria, fungus, protozoa) has certain enzymes and complex chemical reactions that help that organism assimilate carbon. As waste products are generated and the original organic residues are decomposed, new microorganisms may take over, feeding on the waste products, the new flourishing microbial community (generally bacteria), or the more resistant SOM. The early decomposers generally attack the easily digested sugars and proteins followed by microorganisms that attack the more resistant residues.

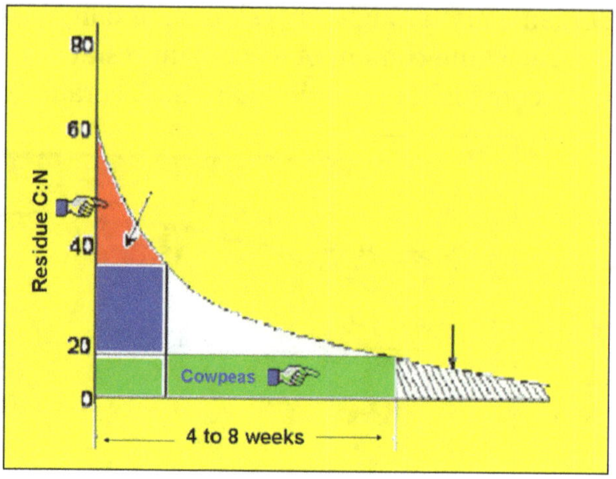

Decomposition of Cover Crop Residues: Cowpeas with a low C:N ratio (<20) will decompose in four to eight weeks and result in net mineralization or release of N. Sudan grass or cereal rye with a higher C:N ratio (>38) will decompose slowly (three months to one year or more) and will result in net immobilization or will tie up soil N.

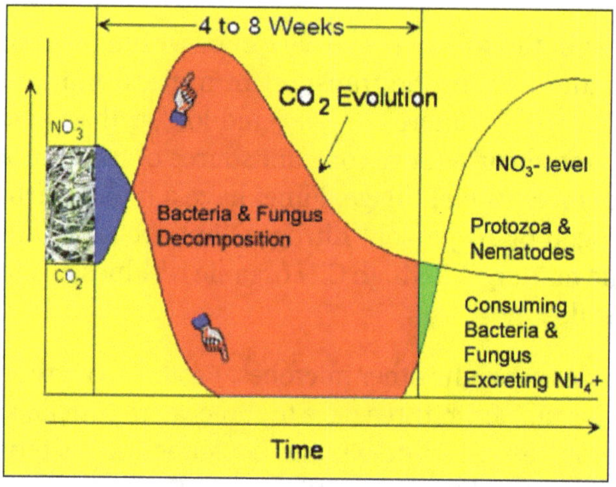

Graph of Cowpeas (C:N<20) being decomposed by bacteria and fungus, the carbon dioxide evolution and protozoa and nematodes consuming the bacteria and fungus and excreting ammonium into the soil for plant growth. NO_3^- and NH_4^+ are easily converted in the soil.

Cover crops supply food (active carbon like glucose and proteins) to the microbes to feed on. In the soil, there are 1,000 to 2,000 times more microbes associated with roots than are living in bare or tilled soil. The microbes in turn build SOM and store soil nutrients. Building SOM requires soil nutrients like N-P-K-S to be tied up in the soil. Winter cover crops soak up excess soil nutrients and supply food to all the microbes in the soil during the winter months rather than microbes having to use up SOM reserves for nutrients. In a conventional tilled field, soil nutrients are quickly released as SOM is burned up and the microbes and soil organisms habitat are destroyed. In a no-till field, high levels of SOM are reserves of soil nutrients which are slowly released into the soils. Adding a living cover crop to a no-till field increases active organic matter (sugars and proteins) for the soil microbes. Soil microbes have two crops to feed on instead of one crop per year. Microbes thrive under no-till conditions and winter cover crops. Cover crops and manure can be used to feed soil microbes and recycle soil nutrients. As soil microbes decompose organic residues, they slowly release nutrients back into the soil for the winter cover crops or for the preceding crop. Cover crops prevent the nutrients from being lost through soil erosion, leaching, volatilization, or denitrification.

Plant-microbe Ecology

Plants have recently been recognized as a metaorganism that possesses a distinct microbiome and close symbiotic relationships with associated microorganisms. Plant ecology is affected by complex interactions with plant-associated microbes. The roles of both plant-associated microbes and the host in ecosystem function have been recognized, but the detailed mechanisms are unclear. Since plants are immobile, they have coevolved with microbes and acquired a number of mechanisms that modulate the outcome of their interactions. Roots can continuously synthesize, accumulate, and secrete a wide range of compounds into the soil, which are known as the root exudates. Root exudates contain enzymes, water, H+ ions, mucilage, and carbon-containing primary and secondary compounds. Campbell and Greaves observed that the density of microbes in the rhizosphere was 100 times greater than that in the bulk soil. Recent studies have showed that plant root exudates shape the soil bacterial community. A plant species selects a specific rhizosphere bacterial community. The change in the microbial composition generates feedback on the plant relative performance that defines the long-term effects of the soil microbes on their coexistence with that plant species. The feedback can be of two types; positive plant-soil microbial feedback reinforces the spatial separation of the microbial communities, while negative feedback results in plant replacement, which necessitates recolonization of locally specific roots. Systematic methods such as genome-wide association studies have enabled us to explore the relationships of plant loci and symbiotic communities in detail. How does the microbiome diversity and function potentially affect host plant performance? The presence of microbial hubs in plant microbiome networks plays an important role between a plant and its microbial community. Plant growth–promoting rhizobacteria (PGPR) can produce a complex blend of volatile substances, which are distinct between bacterial species and other closely related species. Some of these bacterial volatiles can stimulate plant growth, suppress disease stimulating ISR, or antagonize phytopathogens, nematodes, or insects.

Worldwide crop production is affected by biotic and abiotic stress factors, which cause millions of dollars in losses. The beginning of the industrial revolution in 1750 and the human activities such as burning fossil fuels and deforestation have altered the global climate. An increase in carbon dioxide and temperature is speeding up the life cycle of grain and oilseed crops. According to Lesk, extreme heat waves and droughts have reduced global harvests of cereals such as maize, wheat, and rice by 10% in a span of 50 years, which has become a grave concern of various governments. The impact of a warming climate on spring plant phenology is evident. A longer growing season may increase carbon uptake and potentially mitigate climate change, leaf emergence, fruiting, and germination. Abiotic stress factors include extreme temperature, drought, water logging, light, and salinity as major parameters that affect plant growth. Plant-associated microbes or PGPR were found to benefit plants by enhancing nutrient uptake, stimulating root, and shoot growth by producing indole acetic acid, 1-aminocyclopropane-1-carboxylate (ACC) deaminases, solubilizing phosphate, and enhancing the uptake of nutrients from the environment. They also assist and enhance plant resistance to adverse environmental stresses such as drought, heavy metals, salts, and nutrient deficiency. Biotic stress factors include interaction with other organisms and infection by pathogens or damage by insect pests, and some plant growth–promoting bacteria have been used as biocontrol agents against plant pathogens.

Plant Community Ecology and Soil Symbiont Interactions

Plant ecology can be affected by global climate change in terms of above- and below-ground ecological diversity in a terrestrial ecosystem. The terrestrial plant community drives dynamic changes in the soil microbial ecology that may result in alterations in ecosystem function. The ecologist Dr. Peter Chesson postulated that stabilizing mechanisms are essential for maintenance of species diversity and coexistence. Bever et al. have found evidence that microbially mediated positive and negative feedback might play a crucial role in the entire plant ecosystem and contribute to these mechanisms of plant-plant interactions. Traditionally, competing plant species have been thought to have strong negative intraspecific interactions for the high overlap in resource usage. However, success in finding a mechanism to explain the coexistence of competing plant species in maintaining local diversity. Neglecting soil microbial community feedback might be one reason.

Soil Resources and Microbial Interactions

Soil resources can govern the coexistence of plant species by resource partitioning and sharing. Studies have found root symbionts that increase the efficiency of nutrient uptake and allow the host to persist in a low nutrient environment, thereby directly contributing to the competitive exclusion of other plants. Rhizosphere microbes can alter the availability of different forms of nitrogen or phosphorus in the soil and affect plant-plant interactions via the mediation of resource partitioning. Soil resources can also be transferred by shared symbiotic fungi called common mycorrhizal networks (CMNs). In nature, different plant species commonly share the broadly specific mycorrhizal fungi. Simard and Durall demonstrated the direct transfer of resources from one plant to another via CMNs with labeled carbon, nitrogen, and phosphorus. Plant community dynamics are driven by the microbial mediation of soil resource partitioning and sharing.

Host Response to Microbes and Soil Community Feedback

The dynamic density and composition of the rhizospheric microbes can affect the coexistence of plant species via indirect feedback (i.e., the competition or inhibition of symbionts) in the plant population. Ecologists have proposed three hypotheses to explain the mechanism that produces low diversity plant communities. The empty niche hypothesis suggests that novel symbionts inhabit the areas invaded by invasive plants. These symbionts are more efficient at resource acquisition and preferentially associated with invasive plants than with other plants. The degraded mutualist hypothesis suggests that invasive plants and their symbionts inhibit the ability of the native symbiotic community to acquire resources, indirectly reducing the performance of native plants. Positive feedback might be exemplified by the enhanced growth and survival of exotic seedlings near native established symbionts. Plant monodominance, coexistence, and invasion ecology have high relation to symbiotic microbial interactions.

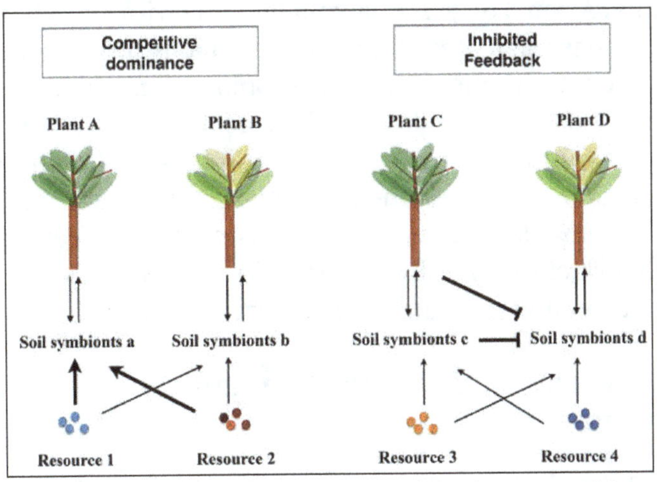

Plant soil community feedback in low diversity communities.

Plants Shape the Microbial Community

Microbial interactions play a crucial role in plant community ecology and performance. How do plants harbor unique microbial communities? How do plants shape a unique rhizosphere microbial community? These are the questions that must be addressed. Modern genomic technologies (e.g., high throughput sequencing) can provide clues to the answer. Lundberg et al. attempted to define the core Arabidopsis thaliana root microbiome. They collected more than 600 A. thaliana plants and compared the bacterial communities using 16S rRNA gene sequencing. They observed that the root microbial communities of plants are sufficiently dependent on the host genotype to vary between inbred A. accessions. However, the mechanisms were not clear but included differences in the host physiology and immune responses.

Plant Genes Responsible for Defense affect the Variation of the Microbial Community

Several studies showed that plant genotype has a small but significant effect on the composition of the endophytic, rhizosphere, or phyllosphere microbial communities. A quantitative trait locus (QTL) analysis and a genome-wide association study (GWAS) were used to identify taxa linked to host genes in humans, mice, plants, and flies. A GWAS of the A. thaliana leaf microbial community suggested that the A. thaliana loci are responsible for defense and that cell wall integrity affected the composition of this community. Furthermore, host genetic variation shaped species richness in the bacterial community. Horton et al. showed that defense response associated genes against pathogens shaped this microbial ecosystem. In Matthew's study, 196 accessions of A. thaliana were sown in a greenhouse and transferred to a field site. The field experiment data suggested that the plant tissue structure (i.e., the cell wall integrity) might affect the leaf microbial community. To understand the plant host genetic factors that affect the associated microbial population, Bodenhausen et al. used a candidate gene approach to investigate the host effects on the composition and abundance of the A. thaliana phyllosphere community. A panel of 55 A. thaliana mutants with alterations in the cell wall, surface structure, defense signaling, secondary metabolism, and pathogen recognition was constructed to reveal the effect on the microbiota composition and/or abundance in a small number of single host mutations. The results showed that lacs and pec1 mutants affected cuticle formation, which led to an increased bacterial abundance and community composition. Moreover, the ethylene signaling gene ein2 was observed to be a host factor that modulated the community composition. Peiffer et al. also noted that approximately 19% of the interline variation in species richness could be attributed to the host genotype in different maize lines grown in the same geographical regions.

Role of Root Exudates in Shaping Rhizosphere Microbial Community

More than a century ago, Lorenz Hiltner defined the term "rhizosphere" as the soil compartment affected by plant roots. Soil microbes are chemotactically attracted to plant root exudates, volatile organic carbon, and rhizodeposition, and then proliferate in this carbon-rich environment. Plant root exudates differ between plant species, so differences in rhizosphere microbiomes of different plant species are expected. More recent studies have provided strong evidence for plant species-specific microbiomes. Plants can also shape the microbial community via root exudates. Root exudates can be categorized as sugars, amino acids, organic acids, nucleotides, flavonoids, antimicrobial compounds, and enzymes.

The types of Root Exudate

Organic Acids and Amino Acids

The composition of root exudates from different cultivars affects the growth of soil-borne pathogens. The susceptible peanut cultivar Ganhua-5 (GH) and the mid-resistant cultivar Quanhua-7 (QH) were chosen for a root exudate analysis and evaluated for the responses of the soil-borne pathogens Fusarium oxysporum and Fusarium solani. The contents of total amino acids, alanine, and sugars in the root exudate of susceptible cultivars were significantly higher than in the mid-resistant cultivar, whereas the contents of total phenolic acids, p-hydroxybenzoic acid, benzoic acid, and p-coumaric acid were significantly lower than in mid-resistant cultivars. These differences in the root exudate composition of susceptible and resistant cultivars might be assumed to regulate the resistance mechanism in the peanut rhizosphere. However, the spore germination and mycelial growth of the soil-borne pathogens F. oxysporum and F. solani were significantly enhanced by treatment with the root exudates from both the susceptible and mid-resistant cultivars compared with a control. If root exudates do not directly inhibit the growth of pathogens, the effects of other factors must be considered. Organic acids modulated the colonization and enhanced the biofilm formation of the root microbiome. Yuan et al. demonstrated that organic acids from banana root exudates facilitated the root colonization by Bacillus amyloliquefaciens. Fumaric acid significantly induced biofilm formation, whereas malic acid evoked the greatest chemotactic response. The results showed that organic acids from banana root exudates played a crucial role in attracting and initiating PGPR colonization on the plant roots. Rice exudates that primarily contained the amino acid residues of histidine, proline, valine, alanine, and glycine, and the carbohydrates glucose, arabinose, mannose, galactose, and glucuronic acid may induce a higher chemotactic response by the endophytic bacteria Corynebacterium flavescens and Bacillus pumilus.

Sugars

The amount of sugar secretion might affect infection by plant pathogens. Gou et al. showed that the Arabidopsis vacuolar sugar transporter SWEET2 limited the Glc-derived carbon efflux from roots and inhibited Pythium infection. They proposed that the expression of SWEET2 modulated sugar secretion, limiting the carbon loss to the rhizosphere. The reduction of available substrates in the rhizosphere contributed to the resistance to Pythium.

Antimicrobial Compounds

Root exudates can also participate in belowground plant defense. Low-molecular-weight antimicrobial chemicals can be divided into "phytoanticipins" and "phytoalexins". Phytoanticipins are defensive compounds that are constitutively produced and released by the plant root prior to a biotic stress such as pathogen infection. Arabidopsis roots deficient in diterpene rhizathalene A production were found to be more susceptible to insect herbivory. Therefore, rhizathalene A was considered as a part of a constitutive direct defense system of the roots. Phytoalexins were defined as inducible defensive compounds that are not detected in healthy plants. Five phenylpropanoid root-derived aromatic root exudates were induced by the attack of the soil-borne pathogen Fusarium graminearum and exhibited antifungal activity. In general, root-secreted terpenoid and phenolic defensive compounds have strong antibacterial and antifungal activity. The largest class of plant defensive chemicals above- and below ground is terpenoids. Nonvolatile terpenoids can be secreted into the rhizosphere, and volatile organic compounds (VOCs) can be emitted from the roots as plant defensive compounds. Phenylpropanoids are a group of plant defensive phenolic root exudates. After a Fusarium graminearum infection, barley rapidly accumulated and secreted phenylpropanoids, which are cinnamic acid derivatives to resist a fungal attack. Phenolic root exudates not only have antimicrobial activity but also beneficially attract soil-borne microorganisms that affect the native soil microbial community. Plants can shape the specific rhizosphere microbial community via root exudates.

Environmental Factors effects on Root Exudates

Plants with different genotypes produce root exudates with different compositions. Abiotic and biotic factors also affect root exudates. Physico-chemical soil properties such as nutrient availability, organic matter content, pH, structure, and texture can affect the availability of root exudates and microbial recruitment by the plant roots. Some biotic factors such as soil microbial secondary metabolism can also affect the exudates.

Temperature

Since the onset of climate change and global warming, the resultant extreme heat and cold have affected the harvest of several crops. To elucidate the effects of temperature on root exudates, Husain and McKeen grew strawberry plants at 5–10°C and compared them with plants grown at 20–30°C. They found more amino acids in exudates in plants grown at a low soil temperature that markedly affected the pathogenicity of Rhizoctonia fragariae. Pramanik demonstrated that in Japanese cucumber grown hydroponically in a growth chamber at high and low temperatures, the organic acid content increased with the elevation of temperature, and some of the compounds identified significantly inhibited plant root growth and germination of cucumber.

Soil Moisture

Flood and drought have reduced global cereal harvests. Several reports have demonstrated that the soil moisture affects the release of root exudates. The temporarily wilting of plants increased the release of amino acids from the plant roots, which might be related to the incidence of pathogens in the rhizosphere. Plants such as peas, soybeans, wheat, barley, and tomatoes were grown in normal moist sand and dried, remoistened sand for the liberation of amino acids. The total amount of amino nitrogen in the temporarily dried sand was many times higher than in the normal moist sand.

Soil pH and Nutrition

The soil pH status and the availability of nutrients such as carbon, nitrogen, and phosphate have been found to affect the release of plant root exudates and the creation of specific chemical niches in the soil, as well as the abundance of plant pathogens and beneficial microbes. Bowen first demonstrated a marked effect of nutrient status on the exudation of amides and amino acids from roots of Pinus radiate seedlings. The results indicated a doubling of amides/amino acids in exudates from phosphate deficient plants. Toljander et al. analyzed the community of arbuscular mycorrhizal (AM) fungi in maize in a long-term fertilization trial and indicated the composition of AM fungi and bacteria was significantly affected and correlated with changes in pH, phosphate, and the soil carbon content. Dumbrell et al. surveyed the AM fungal community of 425 individual plants from 28 plant species. The results showed the strong support for the hypothesis that niche differentiation was based on the structuring of the AM fungal community by soil pH. Root secretion of phenolics was induced in Fe-deficient soil and altered the microbial community in the rhizosphere.

Microorganisms

Soil microorganisms play a crucial role in plant growth and plant exudates. Microorganisms can affect exudation by affecting the permeability of root cells and root metabolism. Microorganisms can also absorb certain compounds in root exudates and excrete other compounds. Soil microbes can produce secondary metabolites that affect plant signaling and metabolism and can be considered as a "plant secondary genome" that provides plant hosts with microbe-derived compounds. Some microbes and also some antibiotics (e.g., penicillin and polymyxin) increased the exudation of organic materials, altered cell permeability, and increased leakage. Soil microbes can also induce the exudation of phenolic compounds for enhancing plant Fe absorption in low-Fe availability soil.

Microbial Community Diversity and Plant Performances

Variation of Microbial Community in Plant Life Cycle

Plant and rhizosphere microbial diversity varies throughout the plant life cycle. The factors influencing the composition and diversity of the microbial community can be classified as four processes: Dispersal, drift, speciation, and selection. For seed plants, the life cycle begins with a seed. Seed dispersal is an important ecological process. Seeds carry associated microbes that originate from their parent and the environment, thereby increasing the microbial diversity in a new environment. Recent studies have suggested that bacterial seed coatings can protect against pathogens. Microbial seed epiphytes have an advantage over soil bacteria during plant colonization. Seed coating methods are a major area of research, and numerous patents have been filed (i.e., approximately 4000 results were found by a Google patent search for the key word "microbial seed coating". After seed dispersal, during seed germination, seed-bone microbes might gain a competitive advantage over other microbes to colonize after germination, and opportunistic microbes from the surrounding soil might have access to a novel niche as the plant develops. Microbial diversity and the community dynamically change throughout the plant life cycle.

Networking of Plant-microbes (Hub and Edge Microbes)

Plant microbiota forms a complex network. A wide range of studies has demonstrated that plant-associated microbes live either inside plant tissue or on the surface of plant organs such as the leaves and roots. Agler et al. characterized the microbiome of A. thaliana leaves. Field experiments showed that both plant genotype and abiotic factors affected the microbiome composition. In addition, they observed that specific species (e.g., the plant pathogen Albugo and the fungus Dioszegia) significantly affected the microbial community structure. Agler used the term "microbial hubs" for the presence of these specific species, which were strongly interconnected with other species in the microbial network of the plants. Microbial hubs might be responsible for mediating defense signals among plants and the effectiveness of biological control agents. The term "keystone species" has been proposed for the presence of a kind of hub species that would be a determinant of colonization of widely microbial taxa. These microbial hubs and keystone species have a large impact on plant performance. A number of hypothetical relationships between plant performance and microbial diversity and composition have been proposed.

Microbial hubs might indirectly affect other taxa by changing host performance, response, or metabolites without directly interacting with other microbes. How can the microbial hubs be identified and how can the interaction of plants and microbes be understood? The requirement of new techniques to analyze whether a microbe has successfully entered a plant, and the observation of changes in the genotypic and phenotypic expression will be an added advantage in the study of plant-microbe interactions.

Plant Growth-promoting Microbes

The soil constitutes a pool of microscopic life forms including bacteria, fungi, actinomycetes, protozoa, and algae, and of these, bacteria are by far the most common. The highest numbers of bacteria are found in the rhizosphere, the region around the plant roots, as differentiated from the bulk soil. Regardless of the concentration of bacteria in the soil, the bacteria may affect a plant in one of three ways. From the perspective of the plant, the interaction between the soil bacteria and a plant may be beneficial, harmful, or neutral. Plant growth-promoting bacteria (PGPB) include those that are free living, those that form specific symbiotic relationships with plants (e.g., Rhizobia spp. and Frankia spp.), bacterial endophytes that can colonize some, or a portion of a plant's interior tissues and cyanobacteria. PGPB can promote plant growth directly by facilitating the acquisition of compounds or modulating plant hormone levels and indirectly by reducing the inhibitory effect of pathogenicity and plant growth by acting as biocontrol agents.

PGPB and Abiotic Stress

In nature, all living organisms are affected by environmental factors such as abiotic stress. Some plants have internal mechanisms to cope up with such stress, while others overcome. Abiotic stress factors include water deficit, excessive water, extreme temperatures, and salinity. The association of PGPB with certain plants can help the plants combat certain abiotic stresses and prevent the plants from dying. In the past decade, bacteria belonging to different genera including Rhizobium, Bacillus, Pseudomonas, Pantoea, Paenibacillus, Burkholderia, Achromobacter, Azospirillum, Microbacterium, Methylobacterium, Variovorax, and Enterobacter have been reported to endow host plants under different abiotic stress environments.

Cold Stress

Maize plants exposed to low temperatures show reduced shoot and root growth that has been attributed to severe oxidative damage induced by cold stress. Treatment with Pseudomonas sp. DSMZ 13134, B. amyloliquefaciens subsp. plantarum, Bacillus simplex strain R41 with micronutrients (Zn/Mn), or seaweed extracts proved to be beneficial cold stress protectant. Inoculation of tomato seeds with plant growth–promoting psychrotolerant bacteria from the genera Arthrobacter, Flavobacterium, Flavimonas, Pedobacter, and Pseudomonas significantly improved plant height, root length, and membrane damage in leaf tissues as evidenced by electrolyte leakage and the malondialdehyde content. A cold-tolerant PGPB Methylobacterium phyllosphaerae strain IARI-HHS2-67, isolated using a leaf imprinting method from phyllosphere of wheat (Triticum aestivum L.), showed improved survival, growth, and nutrient uptake compared to a noninoculated control at 60 days under low-temperature conditions. The chilling resistance of grapevine plantlets was enhanced when inoculated with a plant growth–promoting rhizobacteria, Burkholderia phytofirmans strain PsJN. The root growth increased by 11.8- and 10.7-fold at 26 and 4°C, respectively, and plantlet biomass increased by 6- and 2.2-fold at 26 and 4°C, respectively.

Heat Stress

The effects of global warming in recent years can be felt with the increase in global temperature. A thermo tolerant plant growth–promoting Pseudomonas putida strain AKMP7 was proven to be beneficial for the growth of wheat under heat stress. The bacterium significantly increased the root and shoot length and dry biomass of wheat as compared to uninoculated plants. Inoculation improved the level of cellular metabolites and reduced the activity of several antioxidant enzymes and membrane injury. Sorghum seedlings showed enhanced tolerance to increased temperature with the association of Pseudomonas sp. strain AKM-P6. Inoculation induced the biosynthesis of high-molecular-weight proteins in the leaves at elevated temperatures, reduced membrane injury, and improved the levels of cellular metabolites such as proline, chlorophyll, sugars, amino acids, and proteins.

Salinity

Approximately 20–50% of crop yields are lost to drought and high soil salinity. The United Nations Population Fund estimates that the global human population may well reach 10 billion by 2050. Crop plants are very sensitive to soil salinity, and it is one of the harshest environmental factors that limits the productivity of crops. Plant-microbe associations have been found to be beneficial against abiotic salt stress in Zea mays upon coinoculation with Rhizobium, while Pseudomonas was correlated with decreased electrolyte leakage and the maintenance of leaf water content. Salinity resistant Pseudomonas fluorescens, P. aeruginosa, and P. stutzeri ameliorated sodium chloride stress in tomato plants, and an increase in roots and length were observed. Jha et al. demonstrated that the endophytic bacteria Pseudomonas pseudoalcaligenes induced the accumulation of higher concentrations of glycine betain-like compounds, leading to improved salinity stress tolerance in rice. Dietzia natronolimnaea, a plant growth–promoting rhizobacteria, was seen to modulate a stress response gene, which led to the protection of wheat from salinity stress. Staphylococcus saprophyticus ST1 and Oceanobacillus profundus Pmt2 inoculants were able to produce a biofilm and an extracellular EPS, thus helping Lens esculenta Var. Masoor-93 to cope with salt stress. Salt-stressed Arabidopsis plants treated with volatile organic compounds (VOCs) from B. amyloliquefaciens GB03 showed higher biomass production and less Na^+ accumulation compared to salt-stressed plants without VOC treatment.

Water Stress Resistance

Water scarcity constrains plant productivity, and more crop productivity is lost due to water scarcity than any other abiotic stresses. Achromobacter piechaudii ARV8 reduced the production of ethylene by tomato seedlings following water stress, and ARV8 did not affect the reduction of the relative water content during water deprivation. ARV8 significantly improved the recovery of plants when watering was resumed. Water stress resistance was enhanced in green gram when treated with P. fluorescens Pf1 compared to untreated plants. P. fluorescens Pf1 was also found to produce the enzyme catalase under stress conditions, which helped to detoxify the compounds accumulated in green gram during adverse conditions.

Heavy metals are defined as metals with a density higher than 5 g/cm^3. Heavy metals cause a significant decrease in plant growth and protein content at high concentrations. The most common heavy metal contaminants are Hg, Cd, Cr, Cu, Pb, and Zn. All of these elements are toxic to crop plants at high tissue concentrations. Heavy metal toxicity in plants leads to the production of reactive oxygen species that block essential functional groups of biomolecules. This reaction has been noted in Hg and Cd toxicity and causes oxidative injury in plants. Increasing concentrations of Hg (5–20 mg/kg soil) in tomato plant showed deleterious effects on survival percentage, germination, flowering, pollen viability, and reduced plant height. P. putida enhanced the Cd uptake potential of Eruca sativa and favored healthy growth under Cd stress by increasing the shoot length up to 27%, the root length up to 32%, the wet weight up to 40%, the dry weight up to 22%, and the chlorophyll content up to 26%. Canola seeds inoculated with Kluyvera ascorbata SUD165 and grown under gnotobiotic conditions in the presence of high concentrations of nickel chloride were partially protected against nickel toxicity because the bacteria could lower ethylene-induced stress due to nickel toxicity. Photobacterium halotolerans MELD1 facilitated the phytoprotection of Vigna Unguiculata Sesquipedalis against Hg at a concentration of 25 ppm, thus increasing productivity as well as reducing the translocation of Hg to the bean pods. A plant-microbe phytoremediation system was created with the combination of vetiver grass and the functional endophytic bacterium Achromobacter xylosoxidans F3B for the removal of toluene in Ho et al. It was observed that A. xylosoxidans F3B improved the degradation of toluene in vetiver, resulting in a decrease in phytotoxicity and a 30% reduction of evapotranspiration through the leaves. Another study observed that when A. xylosoxidans strain F3B was inoculated in A. thaliana, it helped the plant tolerate a lethal concentration of catechol and phenol and enhanced the phytoremediation and phytoprotection of the plant.

PGPB against Biotic Stress

Plants must withstand adverse abiotic and biotic stresses when they are sessile. Biotic stress in plants mainly includes damage caused by other living organisms such as insects, bacteria, fungi, nematodes, viruses, viroids, and protists. Biotic stress by PGPR can affect plant growth in two different ways; by the direct promotion of plant growth by the production of phytohormones or by facilitating the uptake of certain nutrients. The indirect promotion of plant growth occurs when PGPR lessens or prevents the deleterious effects of phytopathogens. P. fluorescens produces 2,4-diacetyl phloroglucinol, which inhibits the growth of phytopathogenic fungi. Extracellular chitinase and laminarinase were produced by P. stutzeri, which caused the lysis of mycelia of F. solani, which causes root rot. The endophytic B. cenocepacia 869T2 decreased the disease incidence of Fusarium wilt in treated banana plants to 3.4%, compared to 24.5% in noninoculated plants infected in a field test during a 7-month period. The antibiotic Pyrrolnitrin, produced by P. fluorescens BL915 could prevent the damage from Rhizoctonia solani during the damping off of cotton plants.

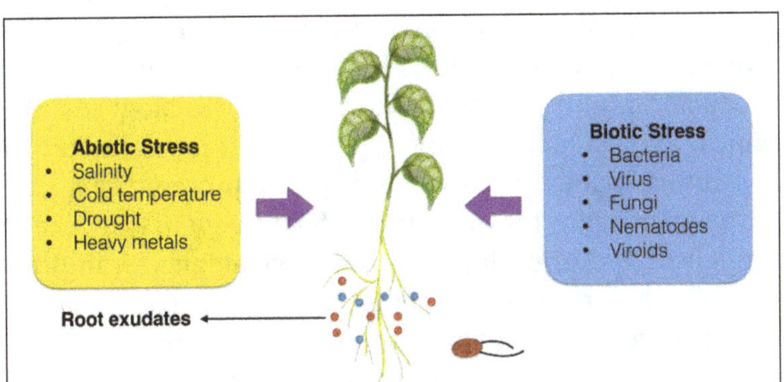

Factors affecting plant-microbe interactions.

Van Peer et al. described a mechanism called "Induced Systemic Resistance" in carnation plants that were systematically protected by P. fluorescens strain WCS417r against F. oxysporum f. sp. Dianthi and by Ardebili et al. in tomato plants, in which P. fluorescens CHA0 protected against F. oxysporum f. sp. Lycopersici acted as a bio agent that induced resistance in tomato. B. amyloliquefaciens strain FZB42, a plant root colonizing isolate, was seen to have the ability to stimulate plant growth and suppress plant pathogens. Endophytes were seen to protect cucumber plants against cucumber anthracnose induced by P. fluorescens strain 89B-61 and Achromobacter sp. F2feb.44. Streptomyces sp. Zapt10, and Bacillus licheniformis AE6 were exploited to induce systemic resistance in cucumber against the foliar disease of downy mildew caused by the phytopathogen Pseudoperonospora cubensis, which enhanced yield.

Pest management has become an issue over time because more and more pests are becoming immune to pesticides. The global pesticide market is growing at a pace of 3.6% per year and is valued around $47 billion. Development of entomopathogenic bacteria for pest management has been a new approach to handle resilient pests. Species belonging to the genera Aschersonia, Agerata, Verticillium, Sphaerostilbe, Podonectria, Myriangium, Hirsutella, and Metarhizium are fungal species involved in the biological control of pests. Bacillus thuringiensis is the most well-known Bacillus species on which the efforts of the scientific community and industry have been focused. Brevibacillus laterosporus Laubach action has been reported to be effective against insects such as Coleoptera, Lepidoptera, nematodes, and phytopathogenic fungi.

Techniques used to understand Plant Microbe Interaction

Plants have been an integral part of our diet since humans began to farm and were no longer nomads. Since then, the world has faced a continuous challenge to feed the ever-growing population. The excess use of fertilizer is causing eutrophication, and genetic engineering of plants is an expensive and tedious process. The role of microbes in plant-microbe interactions has been studied in detail during the past decade. The results of extensive studies suggest that the exploitation of beneficial microbes is a better strategy in the long run to increase crop yield, which will play an important role in disease dissemination and control. Three types of plant-microbe interactions have been studied–symbiosis between plants and mycorrhizae, between plants and rhizobacteria, and pathogenesis. Omics technologies such as transcriptomics, proteomics, and metabolomics coupled with bioinformatics have been extensively applied in plant abiotic stress studies. The proteomics approach has been largely adopted to investigate the protein profiles in plants in response to abiotic stresses that might lead to the development of new strategies for improving stress tolerance. Microbial metabolomics is another technique that has been used to study the set of metabolites present in microbial communities. A rhizosphere metabolomics–driven approach was used in the study of plant-microbe interactions for the removal of polychlorinated biphenyls.

Pathogens and their emergence have been a great threat to food security, agricultural practices, and the conservation of food species, and it has become a significant task to understand the emergence of new pathogens and their role. In the past, plant pathosystems were best studied one gene at a time or one protein at a time; however, the genomic era marked the beginning of the in-depth study of plant-pathogen associations. Commonly known pathogen genomes have been sequenced, and the analysis of these sequences has revealed the forces that have shaped pathogen evolution and has brought to light the unexpected aspects of pathogen biology. The genome scale reconstruction model (GSRM) is based on metabolic reconstructions on a genomic scale for the analysis and interpretation of metabolite concentrations under specific conditions. GSRM has been successfully developed for many organisms including plants, bacteria, fungi, and animals.

Various techniques to study plant-microbe interactions are sequencing, chromatography, mass spectrometry, phospholipid fatty acid (PLFA), microscopy, Fourier transform infrared spectroscopy (FTIR), nuclear magnetic resonance, and real-time PCR (RT-PCR). Further advances in the postgenomic era will pave the way for a better understanding of the interactions of endophytes, plant-pathogen, and plant immunity. Genomic tools to understand major units of the host-microbiota ecology are shown in figure.

References

- Soil-microbiology, land-use-soil: enviroliteracy.org, Retrieved 02 July, 2020
- Scope-and-importance-of-soil-microbiology-135: agriinfo.in, Retrieved 27 May, 2020
- Adesemoye, A. O., Torbert, H. A., and Kloepper, J. W. (2009). Plant growth-promoting rhizobacteria allow reduced application rates of chemical fertilizers. Microb. Ecol. 58, 921–929. doi: 10.1007/s00248-009-9531-y
- Allison, V. J., Condron, L. M., Peltzer, D. A., Richardson, S. J., and Turner, B. L. (2007). Changes in enzyme activities and soil microbial community composition along carbon and nutrient gradients at the Franz Josef chronosequence, New Zealand. Soil Biol. Biochem. 39, 1770–1781. doi: 10.1016/j.soilbio.2007.02.006
- Building Soils for Better Crops, 2nd edition. Sustainable Agriculture Network. Beltsville, MD

Basic Concepts of Soil Microbiology 2

- **Soil Organic Matter**
- **Organic Matter and Microorganisms in Soil**
- **Nutrient Cycle**
- **Nitrogen Fixation**
- **Nitrification**
- **Denitrification**
- **Commercialization of Agriculturally Important Microorganisms**
- **Activity and Variety of Soil Microorganisms**

Some of the basic concepts of soil microbiology include sustainable agriculture, soil organic matter, carbon and phosphorus cycle, nitrogen fixation, nitrification and denitrification, etc. This chapter has been carefully written to provide an easy understanding of the concepts related to soil microbiology.

Soil Organic Matter

Soil organic matter (SOM) is the organic matter component of soil, consisting of plant and animal detritus at various stages of decomposition, cells and tissues of soil microbes, and substances that soil microbes synthesize. SOM provides numerous benefits to the physical and chemical properties of soil and its capacity to provide regulatory ecosystem services. SOM is especially critical for soil functions and quality.

The benefits of SOM result from a number of complex, interactive, edaphic factors; a non-exhaustive list of these benefits to soil function includes improvement of soil structure, aggregation, water retention, soil biodiversity, absorption and retention of pollutants, buffering capacity, and the cycling and storage of plant nutrients. SOM increases soil fertility by providing cation exchange sites and being a reserve of plant nutrients, especially nitrogen (N), phosphorus (P), and sulfur (S), along with micronutrients, which the mineralization of SOM slowly releases. As such, the amount of SOM and soil fertility are significantly correlated.

SOM also acts as a major sink and source of soil carbon (C). Although the C content of SOM varies considerably, SOM is ordinarily estimated to contain 58% C, and "soil organic carbon" (SOC) is often used as a synonym for SOM, with measured SOC content often serving as a proxy for SOM. Soil represents one of the largest C sinks on Earth and is significant in the global carbon cycle. Therefore, SOM/SOC dynamics and the capacity of soils to provide the ecosystem service of carbon sequestration through SOM management have received considerable attention recently.

The concentration of SOM in soils generally ranges from 1% to 6% of the total mass of topsoil for most upland soils. Soils whose upper horizons consist of less than 1% of organic matter are mostly limited to deserts, while the SOM content of soils in low lying, wet areas can be as great as 90%. Soils containing 12% to 18% SOC are generally classified as organic soils.

It can be divided into 3 genera: The living biomass of microbes, fresh and partially decomposed detritus, and humus. Surface plant litter, i. e., fresh vegetal detritus, is generally excluded from SOM.

The primary source of SOM is vegetal detritus. In forests and prairies, for example, different organisms decompose the fresh detritus into simpler compounds. This involves several stages, the first being mostly mechanical, and becoming more chemical as decomposition progresses. The microbial decomposers are included in the SOM, and form a food web of organisms that prey upon each other and subsequently become prey.

There are also other herbivores that consume fresh vegetal matter, the residue of which then passes to the soil. The products of the metabolisms of these organisms are the secondary sources of SOM, which also includes their corpses. Some animals, like earthworms, ants, and centipedes contribute to both vertical and horizontal translocation of organic matter.

Additional sources of SOM include plant root exudates and charcoal.

Composition of Vegetal Detritus

The water content of most vegetal detritus is in the range of 60% to 90%. The dry matter consists of complex organic matter that is composed primarily of carbon, oxygen, and hydrogen. Although these three elements make up about 92% of the dry weight of the organic matter in soil, other elements are very important for the nutrition of plants, including nitrogen, phosphorus, potassium, sulfur, calcium, magnesium, and many micronutrients.

Organic compounds in vegetal detritus include:

- Carbohydrates that are composed of carbon, hydrogen, and oxygen, and range in complexity from rather simple sugars to the large molecules of cellulose.
- Fats that are composed of glycerids of fatty acids, like butyric, stearic, and oleic. They also include carbon, oxygen, and hydrogen.
- Lignins that are complex compounds, form the older parts of wood, and also are composed primarily of carbon, oxygen, and hydrogen. They are resistant to decomposition.
- Proteins that include nitrogen in addition to carbon, hydrogen, and oxygen; and small amounts of sulfur, iron, and phosphorus.
- Charcoal, which is elemental carbon that is derived from incomplete combustion of organic matter. It is resistant to decomposition.

Decomposition

Vegetal detritus in general is not soluble in water and therefore is inaccessible to plants. It constitutes, nevertheless, the raw matter from which plant nutrients derive. Soil microbes decompose it through enzymatic biochemical processes, obtain the necessary energy from the same matter, and produce the mineral compounds that plant roots are apt to absorb. The decomposition of organic compounds specifically into mineral, i. e., inorganic, compounds is denominated "mineralization". A portion of organic matter is not mineralized and instead decomposed into stable organic matter that is denominated "humus".

The decomposition of organic compounds occurs at very different rates, depending on the nature of the compound. The ranking, from fast to slow rates, is:

- Sugars, starches, and simple proteins.
- Proteins.
- Hemicelluloses.
- Cellulose.
- Lignins and fats.

The reactions that occur can be included in one of 3 genera:

- Enzymatic oxidation that produces carbon dioxide, water, and heat. It affects the majority of the matter.
- A series of specific reactions liberates and mineralizes the essential elements nitrogen, phosphorus, and sulfur.
- Compounds that are resistant to microbial action are formed by modification of the original compounds or by microbial synthesis of new ones to produce humus.

The mineral products are:

Element	Mineral Products
Carbon	$CO_2, CO_3^{2-}, HCO_3^-, CH_4, C$
Nitrogen	$NH_4^+, NO_2^-, NO_3^-, N_2$ (gas), N_2O (gas)
Sulfur	$S, H_2S, SO_3^{2-}, SO_4^{2-}, CS_2$
Phosphorus	$H_2PO_4^-, HPO_4^{2-}$
Others	H_2O, O_2, H_2, H^+, OH^-, K^+, Ca^{2+}, Mg^{2+}, etc.

Humus

As vegetal detritus decomposes, some microbially resistant compounds are formed, including modified lignins, oils, fats, and waxes. Secondly, some new compounds are synthesized, like polysaccharides and polyuronids. These compounds are the basis of humus. New reactions occur between these compounds and some proteins and other products that contain nitrogen, thus incorporating nitrogen and avoiding its mineralization. Other nutrients are also protected in this way from mineralization.

Humic Acids/Substances

Humic acids/substances are classified into 3 genera based on their solubility in acids and alkalis, and also according to their stability:

- Fulvic acid is the genus that contains the matter that has the lowest molecular weight, is soluble in acids and alkalis, and is susceptible to microbial action.
- Humic acid is the genus that contains the intermediate matter that has medial molecular weight, is soluble in alkalis and insoluble in acids, and has some resistance to microbial action.
- Humin is the genus that contains the matter that has the greatest molecular weight, is the darkest in color, is insoluble in acids and alkalis, and has the greatest resistance to microbial action.

Function in Carbon Cycling

Soil has a crucial function in the global carbon cycle, with the global soil carbon pool estimated to be 2,500 gigatons. This is 3.3 times the amount of the atmospheric pool at 750 gigatons and 4.5 times the biotic pool at 560 gigatons. The pool of organic carbon, which occurs primarily in the form of SOM, accounts for approximately 1,550 gigatons of the total global carbon pool, with soil inorganic carbon (SIC) accounting for the remainder. The pool of organic carbon exists in dynamic equilibrium between gains and losses; soil may therefore serve as either a sink or source of carbon, through sequestration or greenhouse gas emissions, respectively, depending on exogenous factors.

Organic Matter and Microorganisms in Soil

The most important function of soil microorganisms is the decomposition of various kinds of organic matter present in the soil. Virtually all types of organic matters eventually find their way to the soil or to the sea. The soil organic matter chiefly consists of residues of dead plant and animals, and the excretory products of the living beings.

These organic constituents need to be converted into simple inorganic forms (minerals) to make them available to the autotrophic organisms. This conversion of organic matter into simple inorganic forms is called mineralization.

The mineralization is rendered mainly through decomposition of organic matter by soil microorganisms, mainly fungi and bacteria. It is estimated that 90% of the mineralization of organic matter is the result of the metabolic activities of these two groups of microorganisms.

The remaining 10% results from the metabolism of all other organisms, as well as the combustion of fuel and other materials. The overwhelming contribution of microorganisms to the process of mineralization reflects their ubiquity, their significant contribution to the bulk of living material (their biomass), their high rates of growth and metabolism, and their collective ability to degrade a vast variety of naturally occurring organic materials.

In general, the organic compounds of organism-residues are attacked by organotrophic microorganisms capable of digesting and oxidizing them.

As the oxygen is consumed, anaerobic microorganisms capable of reducing nitrates and sulphates or carbonates commence their activities ultimately creating an aerobic condition. The various simple inorganic substances such as ammonia, sulphide, hydrogen, etc. are avidly oxidized by lithotrophs.

Thus, the organic compounds of organism-residues are completely decomposed to simple organic forms and subsequently oxidized. This cycle of events occur where environment is suitable. Extreme cold and heat are unsuitable for normal turnover of the matter. In cooler parts, decomposition rate is accelerated in spring, whereas in warmer parts rainy conditions favour the activity.

The organic residues which are added to the soil can be categorized into three groups; the easily degradable, moderately degradable and difficultly degradable. These three types of organic residues are attached by distinctly different types of microbes.

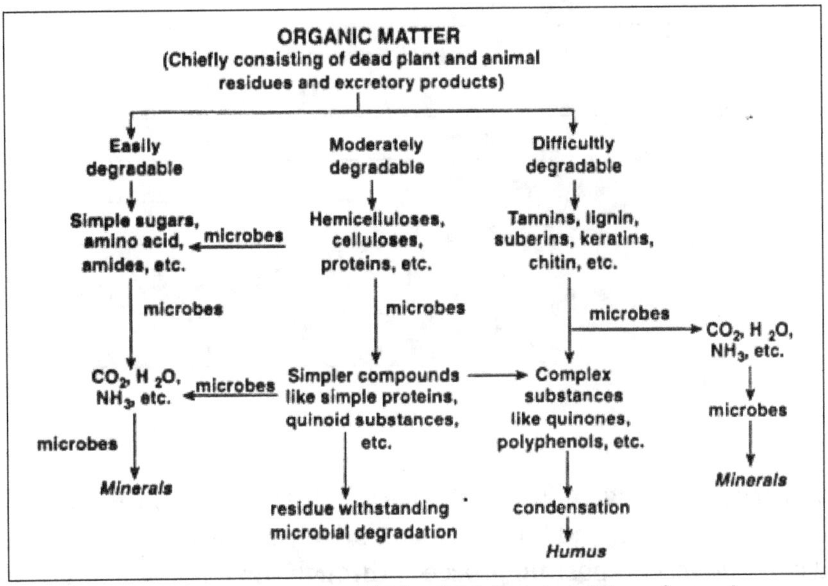

Schematic representation of microbial decomposition of organic matter (mineralization) terminating to huminification.

The result of microbial mineralization is the release of energy, water, gases, etc. on the one hand and formation of a complex amorphous substance (humus, process called humification) on the other hand. The whole process of mineralization terminating into humification is outlined in most simplified way.

The rate of oxidation of biologically important elements such as nitrogen, carbon, sulphur, phosphorus, etc. during mineralization of organic matter is of paramount importance to plants. This is so because these elements, though physically present in large quantities, may not be available to plants as they are not found in suitable chemical form.

These are the microorganisms which bring about these elements to the plants for their use through their activities and also make them free for reuse.

This process of 'use and reuse' through regular recycling of essential elements by the activity of microorganisms during the process of decomposition or mineralization of organic matter is called 'biogeochemical process' which is essential for the maintenance of life on our planet. Different biogeochemical processes, often called biogeochemical cycles.

Thus, the organic matter added to the soil is converted by oxidative decomposition to simpler inorganic substances as a result of mineralization that are made available in stages for plant growth, and the residue transformed into humus as a result of the process called 'humification'; the processes of mineralization and humification running hand-in-hand in nature.

Microorganisms in Compost

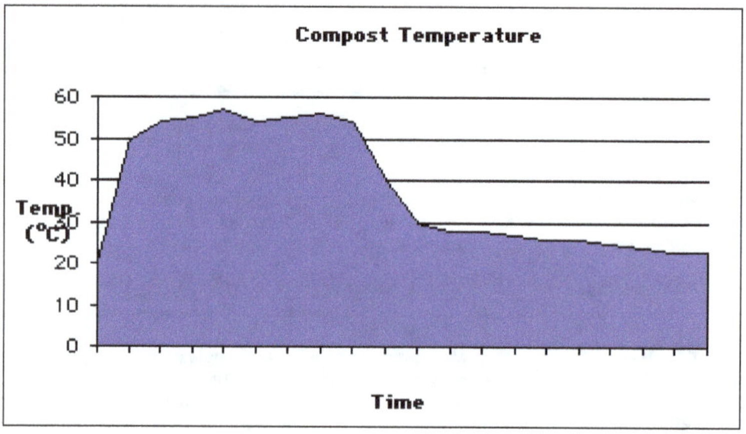

In the process of composting, microorganisms break down organic matter and produce carbon dioxide, water, heat, and humus, the relatively stable organic end product. Under optimal conditions, composting proceeds through three phases: 1) The mesophilic, or moderate-temperature phase, which lasts for a couple of days, 2) The thermophilic, or high-temperature phase, which can last from a few days to several months, and finally, 3) A several-month cooling and maturation phase.

Different communities of microorganisms predominate during the various composting phases. Initial decomposition is carried out by mesophilic microorganisms, which rapidly break down the soluble, readily degradable compounds. The heat they produce causes the compost temperature to rapidly rise.

As the temperature rises above about 40 °C, the mesophilic microorganisms become less competitive and are replaced by others that are thermophilic, or heat-loving. At temperatures of 55 °C and above, many microorganisms that are human or plant pathogens are destroyed. Because temperatures over about 65 °C kill many forms of microbes and limit the rate of decomposition, compost managers use aeration and mixing to keep the temperature below this point.

During the thermophilic phase, high temperatures accelerate the breakdown of proteins, fats, and complex carboydrates like cellulose and hemicellulose, the major structural molecules in plants. As the supply of these high-energy compounds becomes exhausted, the compost temperature gradually decreases and mesophilic microorganisms once again take over for the final phase of "curing" or maturation of the remaining organic matter.

Bacteria

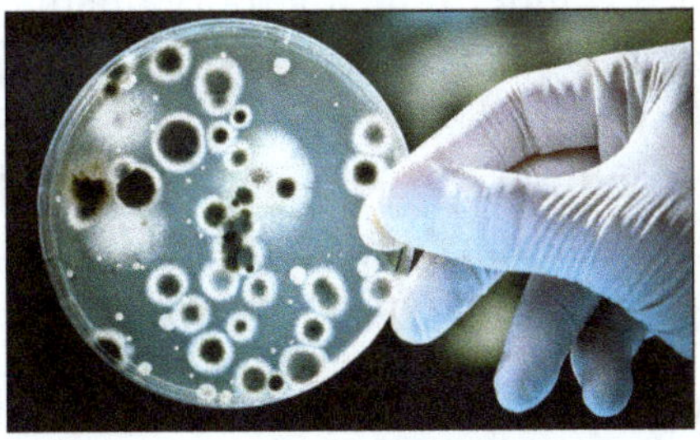

Bacteria are the smallest living organisms and the most numerous in compost; they make up 80 to 90% of the billions of microorganisms typically found in a gram of compost. Bacteria are responsible for most of the decomposition and heat generation in compost. They are the most nutritionally diverse group of compost organisms, using a broad range of enzymes to chemically break down a variety of organic materials.

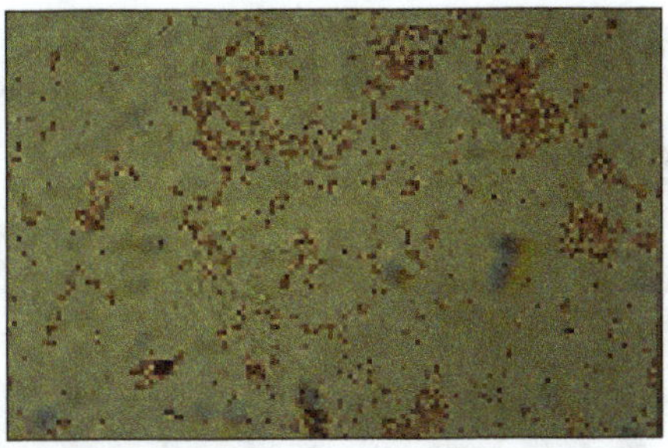

Bacteria are single-celled and structured as either rod-shaped bacilli, sphere-shaped cocci or spiral-shaped spirilla. Many are motile, meaning that they have the ability to move under their own power. At the beginning of the composting process (0-40 °C), mesophilic bacteria predominate. Most of these are forms that can also be found in topsoil.

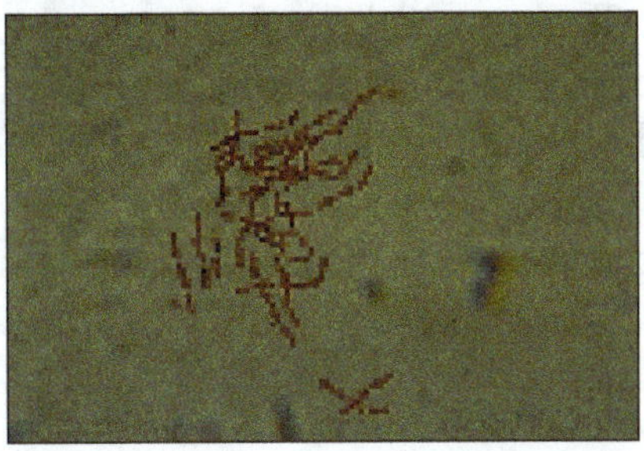

As the compost heats up above 40°C, thermophilic bacteria take over. The microbial populations during this phase are dominated by members of the genus Bacillus. The diversity of bacilli species is fairly high at temperatures from 50-55°C but decreases dramatically at 60°C or above. When conditions become unfavorable, bacilli survive by forming endospores, thick-walled spores that are highly resistant to heat, cold, dryness, or lack of food. They are ubiquitous in nature and become active whenever environmental conditions are favorable.

At the highest compost temperatures, bacteria of the genus Thermus have been isolated. Composters sometimes wonder how microorganisms evolved in nature that can withstand the high temperatures found in active compost. Thermus bacteria were first found in hot springs in Yellowstone National Park and may have evolved there. Other places where thermophilic conditions exist in nature include deep sea thermal vents, manure droppings, and accumulations of decomposing vegetation that have the right conditions to heat up just as they would in a compost pile.

Once the compost cools down, mesophilic bacteria again predominate. The numbers and types of mesophilic microbes that recolonize compost as it matures depend on what spores and organisms are present in the compost as well as in the immediate environment. In general, the longer the curing or maturation phase, the more diverse the microbial community it supports.

Actinomycetes

The characteristic earthy smell of soil is caused by actinomycetes, organisms that resemble fungi but actually are filamentous bacteria. Like other bacteria, they lack nuclei, but they grow multicellular filaments like fungi. In composting they play an important role in degrading complex organics such as cellulose, lignin, chitin, and proteins. Their enzymes enable them to chemically break down tough debris such as woody stems, bark, or newspaper. Some species appear during the thermophilic phase, and others become important during the cooler curing phase, when only the most resistant compounds remain in the last stages of the formation of humus.

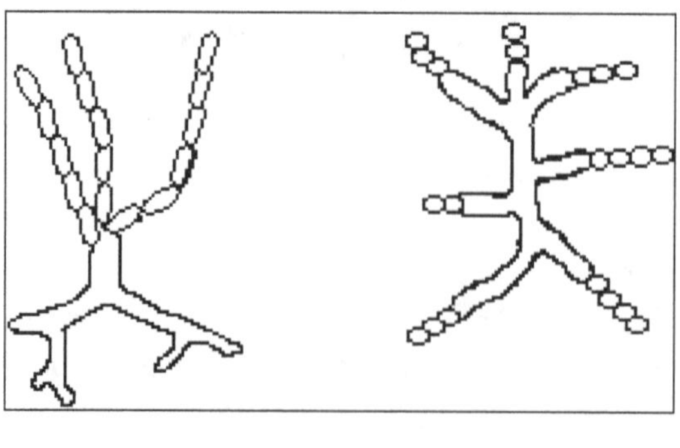

Actinomycetes form long, thread-like branched filaments that look like gray spider webs stretching through compost. These filaments are most commonly seen toward the end of the composting process, in the outer 10 to 15 centimeters of the pile. Sometimes they appear as circular colonies that gradually expand in diameter.

Fungi

Fungi include molds and yeasts, and collectively they are responsible for the decomposition of many complex plant polymers in soil and compost. In compost, fungi are important because they break down tough debris, enabling bacteria to continue the decomposition process once most of the cellulose has been exhausted. They spread and grow vigorously by producing many cells and filaments, and they can attack organic residues that are too dry, acidic, or low in nitrogen for bacterial decomposition.

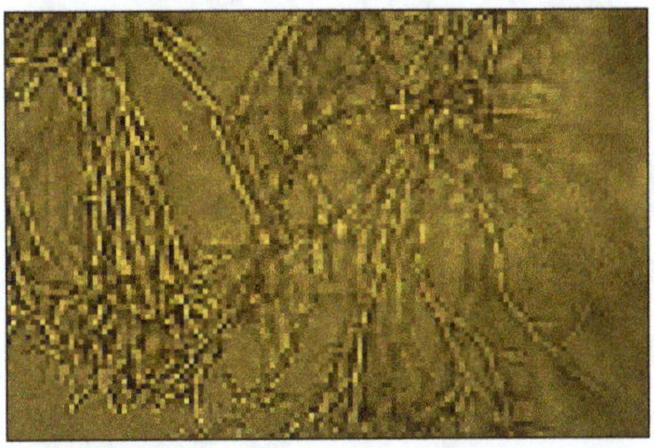

Most fungi are classified as saprophytes because they live on dead or dying material and obtain energy by breaking down organic matter in dead plants and animals. Fungal species are numerous during both mesophilic and thermophilic phases of composting. Most fungi live in the outer layer of compost when temperatures are high. Compost molds are strict aerobes that grow both as unseen filaments and as gray or white fuzzy colonies on the compost surface.

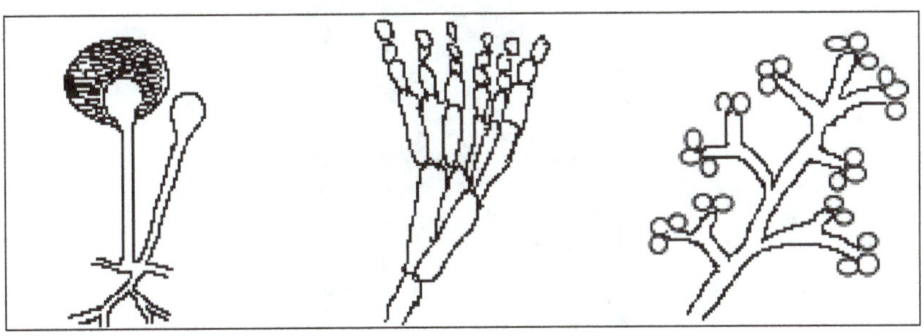

Protozoa

Protozoa are one-celled microscopic animals. They are found in water droplets in compost but play a relatively minor role in decomposition. Protozoa obtain their food from organic matter in the same way as bacteria do but also act as secondary consumers ingesting bacteria and fungi.

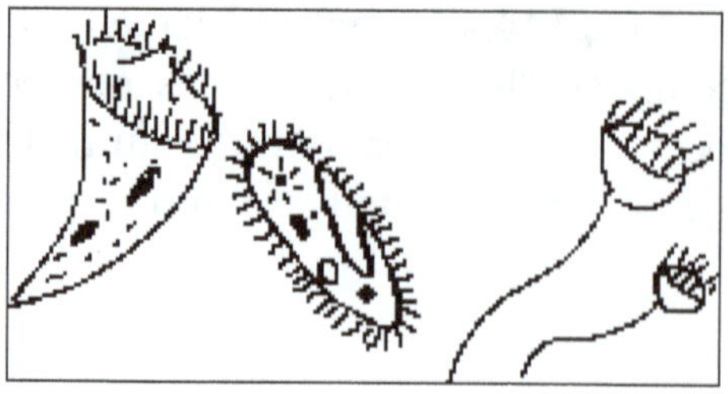

Rotifers

Rotifers are microscopic multicellular organisms also found in films of water in the compost. They feed on organic matter and also ingest bacteria and fungi.

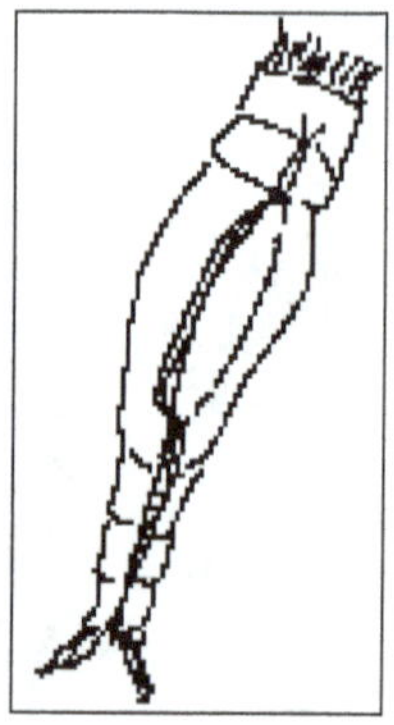

Nutrient Cycle

Fallen logs are critical components of the nutrient cycle in terrestrial forests. Nurse logs form habitats for other creatures that decompose the materials and recycle the nutrients back into production.

A nutrient cycle (or ecological recycling) is the movement and exchange of organic and inorganic matter back into the production of matter. Energy flow is a unidirectional and noncyclic pathway, whereas the movement of mineral nutrients is cyclic. Mineral cycles include the carbon cycle, sulfur cycle, nitrogen cycle, water cycle, phosphorus cycle, oxygen cycle, among others that continually recycle along with other mineral nutrients into productive ecological nutrition.

The nutrient cycle is nature's recycling system. All forms of recycling have feedback loops that use energy in the process of putting material resources back into use. Recycling in ecology is regulated to a large extent during the process of decomposition. Ecosystems employ biodiversity in the food webs that recycle natural materials, such as mineral nutrients, which includes water. Recycling in natural systems is one of the many ecosystem services that sustain and contribute to the well-being of human societies.

There is much overlap between the terms for the biogeochemical cycle and nutrient cycle. However, the terms often appear independently. Nutrient cycle is more often used in direct reference to the idea of an intra-system cycle, where an ecosystem functions as a unit. From a practical point, it does not make sense to assess a terrestrial ecosystem by considering the full column of air above it as well as the great depths of Earth below it. While an ecosystem often has no clear boundary, as a working model it is practical to consider the functional community where the bulk of matter and energy transfer occurs. Nutrient cycling occurs in ecosystems that participate in the "larger biogeochemical cycles of the earth through a system of inputs and outputs."

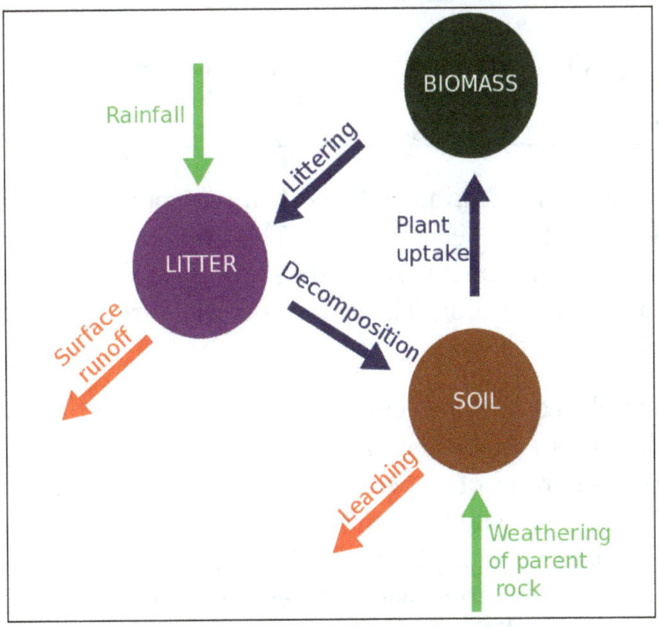

A nutrient cycle of a typical terrestrial ecosystem.

Complete and Closed Loop

Ecosystems are capable of complete recycling. Complete recycling means that 100% of the waste material can be reconstituted indefinitely. This idea was captured by Howard

T. Odum when he penned that "it is thoroughly demonstrated by ecological systems and geological systems that all the chemical elements and many organic substances can be accumulated by living systems from background crustal or oceanic concentrations without limit as to concentration so long as there is available solar or another source of potential energy". In 1979 Nicholas Georgescu-Roegen proposed the fourth law of entropy stating that complete recycling is impossible. Despite Georgescu-Roegen's extensive intellectual contributions to the science of ecological economics, the fourth law has been rejected in line with observations of ecological recycling. However, some authors state that complete recycling is impossible for technological waste.

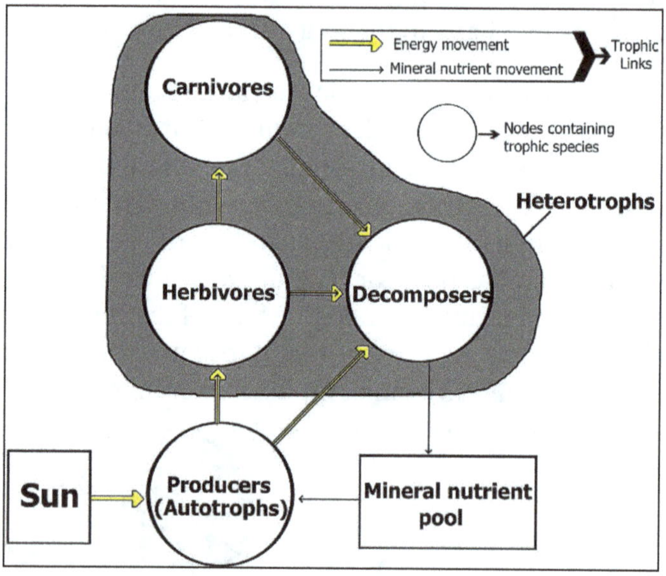

A simplified food web illustrating a three-trophic food chain (*producers-herbivores-carnivores*) linked to decomposers. The movement of mineral nutrients through the food chain, into the mineral nutrient pool, and back into the trophic system illustrates ecological recycling. The movement of energy, in contrast, is unidirectional and noncyclic.

Ecosystems execute closed loop recycling where demand for the nutrients that adds to the growth of biomass exceeds supply within that system. There are regional and spatial differences in the rates of growth and exchange of materials, where some ecosystems may be in nutrient debt (sinks) where others will have extra supply. These differences relate to climate, topography, and geological history leaving behind different sources of parent material. In terms of a food web, a cycle or loop is defined as "a directed sequence of one or more links starting from, and ending at, the same species." An example of this is the microbial food web in the ocean, where "bacteria are exploited, and controlled, by protozoa, including heterotrophic microflagellates which are in turn exploited by ciliates. This grazing activity is accompanied by excretion of substances which are in turn used by the bacteria so that the system more or less operates in a closed circuit."

Ecological Recycling

An example of ecological recycling occurs in the enzymatic digestion of cellulose. "Cellulose, one of the most abundant organic compounds on Earth, is the major polysaccharide in plants where it is part of the cell walls. Cellulose-degrading enzymes participate in the natural, *ecological recycling* of plant material." Different ecosystems can vary in their recycling rates of litter, which creates a complex feedback on factors such as the competitive dominance of certain plant species. Different rates and patterns of ecological recycling leaves a legacy of environmental effects with implications for the future evolution of ecosystems.

Ecological recycling is common in organic farming, where nutrient management is *fundamentally different* compared to agri-business styles of soil management. Organic farms that employ ecosystem recycling to a greater extent support more species (increased levels of biodiversity) and have a different food web structure. Organic agricultural ecosystems rely on the services of biodiversity for the recycling of nutrients through soils instead of relying on the supplementation of synthetic fertilizers. The model for ecological recycling agriculture adheres to the following principals:

- Protection of biodiversity.
- Use of renewable energy.
- Recycling of plant nutrients.

Where produce from an organic farm leaves the farm gate for the market the system becomes an open cycle and nutrients may need to be replaced through alternative methods.

Ecosystem Engineers

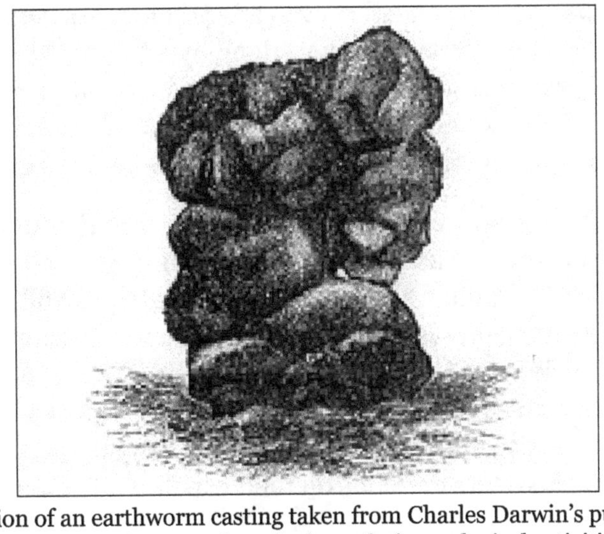

An illustration of an earthworm casting taken from Charles Darwin's publication on the movement of organic matter in soils through the ecological activities of worms.

From the largest to the smallest of creatures, nutrients are recycled by their movement, by their wastes, and by their metabolic activities. This illustration shows an example of the whale pump that cycles nutrients through the layers of the oceanic water column. Whales can migrate to great depths to feed on bottom fish (such as sand lance *Ammodytes* spp.) and surface to feed on krill and plankton at shallower levels. The whale pump enhances growth and productivity in other parts of the ecosystem.

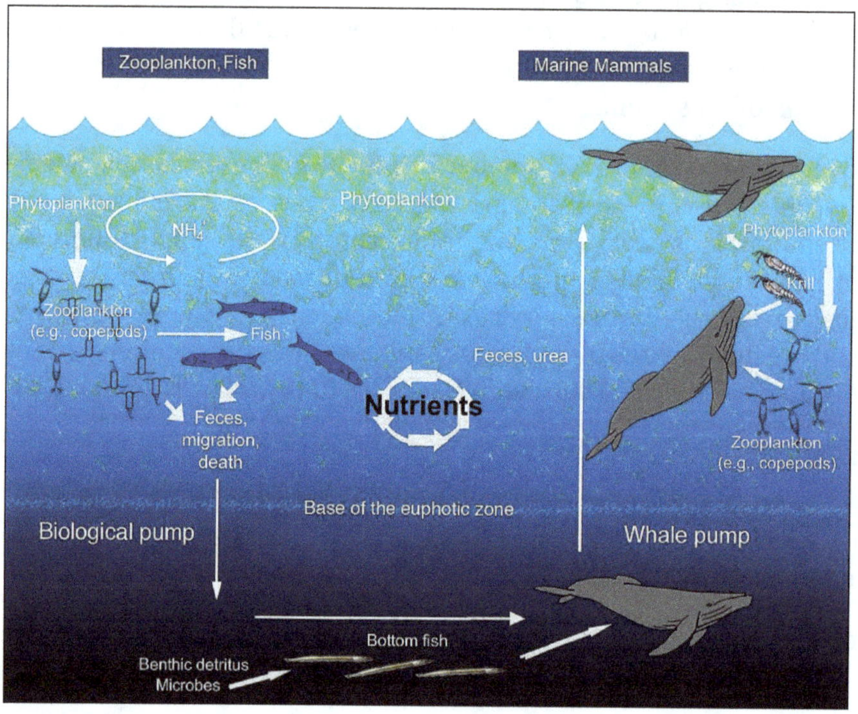

The persistent legacy of environmental feedback that is left behind by or as an extension of the ecological actions of organisms is known as niche construction or ecosystem engineering. Many species leave an effect even after their death, such as coral skeletons or the extensive habitat modifications to a wetland by a beaver, whose components are recycled and re-used by descendants and other species living under a different selective regime through the feedback and agency of these legacy effects. Ecosystem engineers can influence nutrient cycling efficiency rates through their actions.

Earthworms, for example, passively and mechanically alter the nature of soil environments. Bodies of dead worms passively contribute mineral nutrients to the soil. The worms also mechanically modify the physical structure of the soil as they crawl about (bioturbation), digest on the molds of organic matter they pull from the soil litter. These activities transport nutrients into the mineral layers of soil. Worms discard wastes that create worm castings containing undigested materials where bacteria and other decomposers gain access to the nutrients. The earthworm is employed in this process and the production of the ecosystem depends on their capability to create feedback loops in the recycling process.

Shellfish are also ecosystem engineers because they: 1) Filter suspended particles from the water column; 2) Remove excess nutrients from coastal bays through denitrification; 3) Serve as natural coastal buffers, absorbing wave energy and reducing erosion from boat wakes, sea level rise and storms; 4) Provide nursery habitat for fish that are valuable to coastal economies.

Fungi contribute to nutrient cycling and nutritionally rearrange patches of ecosystem creating niches for other organisms. In that way fungi in growing dead wood allow xylophages to grow and develop and xylophages, in turn, affect dead wood, contributing to wood decomposition and nutrient cycling in the forest floor.

Phosphorus Cycle

Phosphorus is one of the most important constituent of several important compounds always present in organisms. It occurs both in organic (nucleic acids, nucleoproteins, phospholipids, etc.) and inorganic (phosphate) forms in the living organisms. Animals possessing bones have large amount of phosphorus in its inorganic form.

However, phosphorus is added to soil through chemical fertilizers, excrete and organism- residues. Though there is plenty of phosphorus present in the soil in unavailable inorganic forms, most of the plants obtain it only as orthrophosphate ions (soluble inorganic forms). However, mycorrhizae, when present, help the plants in obtaining phosphorus.

The cycle of phosphorus can be well studied under following two heads:

- Mineralization: Conversion of Organic Phosphorus into Insoluble Inorganic Phosphates.

  Many soil microorganisms produce enzymes that attack many of the organic phosphorus compounds in the soil and release inorganic phosphate. This process is comparable to the mineralization of organic nitrogen compounds. The enzymes involved in these reactions are collectively called 'phosphatases' which have a broad range of substrate specificity.

- Solubilization: Conversion of Insoluble Inorganic Phosphates into Soluble Inorganic Phosphates.

  The availability of phosphorus depends on the degree of solubilization by various organic and inorganic acids produced by microorganisms in soil. These are the solubilized form of insoluble inorganic phosphates which are taken in by the plants.

  Fungi, e.g., Aspergillus, Penicillium, Fusarium are the most important of the soil microorganisms which produce substantial amounts of these acids; others are the bacteria, namely, Bacillus, Pseudomonas, Micrococcus, Flavobacterium, etc.

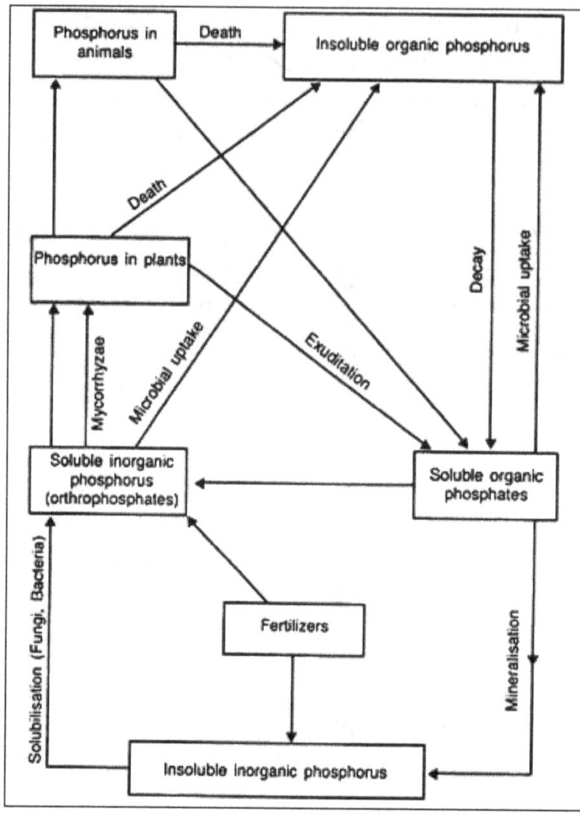

Schematic representation of phosphorous cycle.

The overall conversion of insoluble inorganic phosphates into soluble inorganic phosphates by the action of acids can be exemplified via reactions as under.

(a.i)	$Ca_3(PO_4)_2$ + (Insoluble calcium phosphate)	H_2SO_4 (Sulfuric acid)	⟶	$Ca_2H_2(PO_4)_2$ + (Calcium-monohydrogen phosphate fairly soluble in water)	$CaSO_4$
(ii)	$Ca_2H_2(PO_4)_2$ +	H_2SO_4	⟶	$CaH_4(PO_4)_2$ + (Calcium-dihydrogen phosphate; highly soluble in water)	$CaSO_4$
(b.i)	$Ca_3(PO_4)_2$ + (Insoluble calcium phosphate)	$2HNO_3$ (Nitric acid)	⟶	$Ca_2H_2(PO_4)_2$ + (Calcium-monohydrogen phosphate; fairly soluble in water)	$Ca(NO_3)_2$
(ii)	$Ca_2H_2(PO_4)_2$ +	$2HNO_3$	⟶	$CaH_4(PO_4)_2$ + (Calcium-dihydrogen phosphate; highly soluble in water)	$Ca(NO_3)_2$

The action of acids to convert insoluble phosphates into soluble ones is generally called 'solubilization' and particularly takes place in close proximity of the root surfaces where sugar from root-exudates are converted by the action of microorganisms into organic acids.

Carbon Cycle

Carbon is the most important element in the biological system and constitutes about 50% of all living organisms. Carbon dioxide present in the atmosphere or dissolved in water is the ultimate source of organic carbon compounds occurring in nature.

The cycle of carbon in nature comprises of two main processes:

- The conversion of oxidized form of carbon into reduced organic form by photosynthetic organisms.
- Restoration of original oxidized form through mineralization of the organic form by the micro-organisms.

Conversion of Oxidized form of Carbon (CO_2) into Reduced Organic Form

CO_2 is reduced into organic carbon compounds mainly by the process of photosynthesis. Photosynthetic algae and higher plants are the most important agents of carbon dioxide fixation. In the ocean the major plant forms that fix carbon are the free floating microscopic algae called phytoplanktons. They are estimated to fix annually about 1.2 x 10^{10} tons of carbon.

Nearly 1.6 x 10^{10} tons of carbon is said to be fixed annually by photosynthetic terrestrial plant life. Besides, autotrophic and heterotrophic bacteria are also capable of synthesizing organic matter from inorganic carbon. In addition to the occurrence of photosynthesis among microorganisms, the latter also represent the example of CO_2 fixation into organic compounds which are as follows.

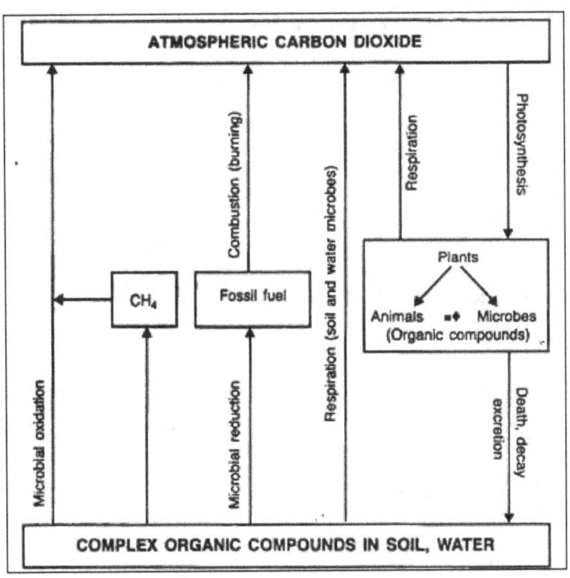

Schematic representation of carbon cycle.

- The carbon dioxide represents the sole source of carbon for autotrophic bacteria. The latter fix CO_2 to carbohydrates by a reduction reaction.

$$CO_2 + 2H_2 \rightarrow (CH_2O)_x + H_2O$$

- Heterotrophic bacteria fix carbon dioxide commonly.

$$\underset{\text{(Pyruvic acid)}}{CH_3COCOOH} + CO_2 \rightarrow \underset{\text{(Oxaloacetic acid)}}{HOOCCH_2 \cdot COCOOH}$$

Restoration of Original Oxidized Form (CO_2) through Mineralization of the Organic Form

One can consider three different modes through which the organic matter is mineralizes and the CO_2 is released in the atmosphere.

They are:

- Process of respiration.
- Accidental (forest fire) and intentional (fuel) burning.
- Decomposition of organic matter by microorganisms.

The process of respiration in plants and animals, and the accidental and intentional burning of plants and their parts result in the breakdown of organic carbon compounds releasing carbon dioxide in the atmosphere.

Decomposition of Organic Matter by Microorganisms

The organic carbon compounds that eventually are deposited in the soil are degraded by the activities of microorganisms which are mainly the bacteria and fungi. The CO_2 is released into the air and soil.

Cellulose Decomposition

Cellulose is the most abundant organic material in plants. It is readily attacked by many species of fungi and bacteria.

The process of cellulose decomposition to carbon dioxide can be summarized in the form of following reactions:

$$\text{Cellulose} \xrightarrow[\text{cellulose}]{\text{enzyme}} Cellobiose$$

$$Cellobiose \xrightarrow[\beta-glu\cos idase]{\text{enzyme}} Glu\cos e$$

$$Glu\cos e \xrightarrow[\text{of many microbes}]{\text{enzyme system}} CO_2 + H_2O + and/or\ other\ end\ products$$

The fungi which decompose cellulose in soil are mainly Trichoderma, Aspergillus, Penicillium, Fusarium, Chaetomium, Verticillium. Rhizoctonia, Myrothecium, Merulius, Pleurotus, Fomes, etc.

The bacteria that bring about cellulose decomposition in soil consist mainly of the species of Clostridium, Cellulomonas, Streptomyces, Cytophaga, Bacillus, Pseudomonas, Nocardia, Micromonospora, Sporocytophaga, Polyangium, Cellfalcicula, etc.

Hemicellulose Decomposition

Hemicelluloses are the polymers of simple sugars such as pentoses, hexoses and uronic acid. The decomposition of hemicelluloses by microorganisms takes place through the agency of extracellular enzymes called hemicellulases.

The fungi that degrade hemicelluloses in soil are examplified by Chaetomium, Aspergillus, Penicillium, Trichoderma, Fusarium, Humicola, etc. Bacillus, Pseudomonas, Cytophaga, Vibrio, Erwinia, Streptomyces, Actinomyces, etc. are the bacteria that degrade hemicelluloses in soil.

Lignin Decomposition

Lignin is the third most abundant constituent of the plants. It is highly resistant of microbial degradation. However, certain fungi (examplified by Aspergillus, Penicillium, Fusarium, Lenzites, Clavaria, Polyporus, etc.) and bacteria (examplified by Streptomyces, Nocardia, Flavabacterium, Xanthomonas, Pseudomonas, Micrococcus, etc.) are known to degrade lignin at slow rates.

Nitrogen Fixation

Nitrogen fixation is a process by which molecular nitrogen in the air is converted into ammonia (NH_3) or related nitrogenous compounds in soil. Atmospheric nitrogen is molecular dinitrogen, a relatively nonreactive molecule that is metabolically useless to all but a few microorganisms. Biological nitrogen fixation converts N_2 into ammonia, which is metabolized by most organisms.

Nitrogen fixation is essential to life because fixed inorganic nitrogen compounds are required for the biosynthesis of all nitrogen-containing organic compounds, such as amino acids and proteins, nucleoside triphosphates and nucleic acids. As part of the nitrogen cycle, it is essential for agriculture and the manufacture of fertilizer. It is also, indirectly, relevant to the manufacture of all nitrogen chemical compounds, which includes some explosives, pharmaceuticals and dyes.

Nitrogen fixation is carried out naturally in soil by microorganisms termed diazotrophs that include bacteria such as *Azotobacter* and archaea. Some nitrogen-fixing bacteria have symbiotic relationships with plant groups, especially legumes. Looser non-symbiotic relationships between diazotrophs and plants are often referred to as associative, as seen in nitrogen fixation on rice roots. Nitrogen fixation occurs between some termites and fungi. It occurs naturally in the air by means of NOx production by lightning.

All biological nitrogen fixation is effected by enzymes called nitrogenases. These enzymes contain iron, often with a second metal, usually molybdenum but sometimes vanadium.

Fixation

Non-biological

Nitrogen can be fixed by lightning that converts nitrogen and oxygen into NO*x* (nitrogen oxides). NO*x* may react with water to make nitrous acid or nitric acid, which seeps into the soil, where it makes nitrate, which is of use to plants. Nitrogen in the atmosphere is highly stable and nonreactive due to the triple bond between atoms in the N_2 molecule. Lightning produces enough energy and heat to break this bond allowing nitrogen atoms to react with oxygen, forming NO_x. These compounds cannot be used by plants, but as this molecule cools, it reacts with oxygen to form NO_2. This molecule in turn reacts with water to produce HNO_3 (nitric acid), or its ion NO_3^- (nitrate), which is usable by plants.

Lightning heats the air around it breaking the bonds of N_2 starting the formation of nitrous acid.

Biological

Biological nitrogen fixation was discovered by German agronomist Hermann Hellriegel and Dutch microbiologist Martinus Beijerinck. Biological nitrogen fixation (BNF) occurs when atmospheric nitrogen is converted to ammonia by a nitrogenase enzyme. The overall reaction for BNF is:

$$N_2 + 16ATP + 8e^- + 8H^+ -> 2NH_3 + H_2 + 16ADP + 16\mathrm{P}_i$$

The process is coupled to the hydrolysis of 16 equivalents of ATP and is accompanied by the co-formation of one equivalent of H_2. The conversion of N_2 into ammonia occurs at a metal cluster called FeMoco, an abbreviation for the iron-molybdenum cofactor. The mechanism proceeds via a series of protonation and reduction steps wherein the

FeMoco active site hydrogenates the N_2 substrate. In free-living diazotrophs, nitrogenase-generated ammonia is assimilated into glutamate through the glutamine synthetase/glutamate synthase pathway. The microbial nif genes required for nitrogen fixation are widely distributed in diverse environments.

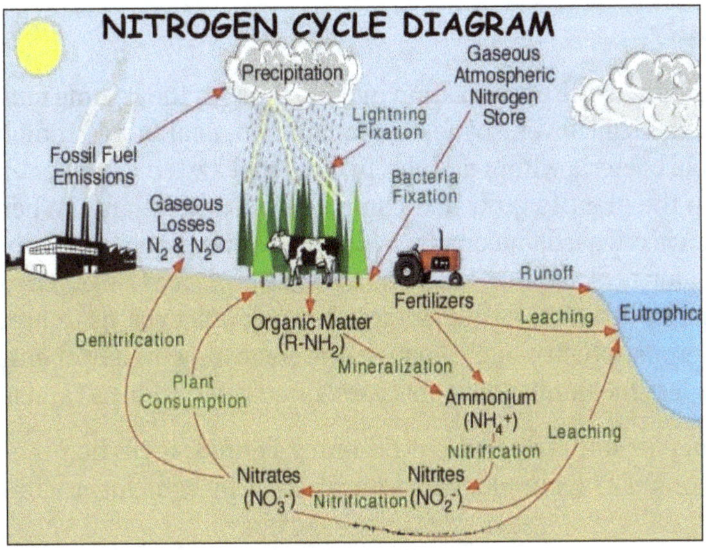

Nitrogen cycle.

Nitrogenases are rapidly degraded by oxygen. For this reason, many bacteria cease production of the enzyme in the presence of oxygen. Many nitrogen-fixing organisms exist only in anaerobic conditions, respiring to draw down oxygen levels, or binding the oxygen with a protein such as leghemoglobin.

Microorganisms

Diazotrophs are widespread within domain Bacteria including cyanobacteria (e.g. the highly significant *Trichodesmium* and *Cyanothece*), as well as green sulfur bacteria, Azotobacteraceae, rhizobia and *Frankia*. Several obligately anaerobic bacteria fix nitrogen including many (but not all) *Clostridium* spp. Some archaea also fix nitrogen, including several methanogenic taxa, which are significant contributors to nitrogen fixation in oxygen-deficient soils.

Cyanobacteria inhabit nearly all illuminated environments on Earth and play key roles in the carbon and nitrogen cycle of the biosphere. In general, cyanobacteria can use various inorganic and organic sources of combined nitrogen, such as nitrate, nitrite, ammonium, urea, or some amino acids. Several cyanobacteria strains are also capable of diazotrophic growth, an ability that may have been present in their last common ancestor in the Archean eon. Nitrogen fixation by cyanobacteria in coral reefs can fix twice as much nitrogen as on land—around 660 kg/ha/year. The colonial marine cyanobacterium *Trichodesmium* is thought to fix nitrogen on such a scale that it accounts for almost half of the nitrogen fixation in marine systems globally.

Marine surface licing and non-photosynthetic bacteria belonging in Proteobacteria and Planctomycetes fixate significant atmospheric nitrogen.

Root Nodule Symbioses

Legume Family

Plants that contribute to nitrogen fixation include those of the legume family—Fabaceae—with taxa such as kudzu, clover, soybean, alfalfa, lupin, peanut and rooibos. They contain symbiotic rhizobia bacteria within nodules in their root systems, producing nitrogen compounds that help the plant to grow and compete with other plants. When the plant dies, the fixed nitrogen is released, making it available to other plants; this helps to fertilize the soil. The great majority of legumes have this association, but a few genera (e.g., *Styphnolobium*) do not. In many traditional and organic farming practices, fields are rotated through various types of crops, which usually include one consisting mainly or entirely of clover or buckwheat (non-legume family Polygonaceae), often referred to as "green manure".

Fixation efficiency in soil is dependent on many factors, including the legume and air and soil conditions. For example, nitrogen fixation by red clover can range from 50-200 lb./acre.

Inga alley farming relies on the leguminous genus *Inga*, a small tropical, tough-leaved, nitrogen-fixing tree.

Non-leguminous

A sectioned alder tree root nodule.

Other nitrogen fixing families include:

- *Parasponia*, a tropical genus in the family Cannabaceae, which are able to interact with rhizobia and form nitrogen-fixing nodules.
- Actinorhizal plants such as alder and bayberry can form nitrogen-fixing nodules, thanks to a symbiotic association with *Frankia* bacteria. These plants belong to 25 genera distributed across eight families.

The ability to fix nitrogen is present in other families that belong to the orders Cucurbitales, Fagales and Rosales, which together with the Fabales form a clade of eurosids.

The ability to fix nitrogen is not universally present in these families. For example, of 122 Rosaceae genera, only four fix nitrogen. Fabales were the first lineage to branch off this nitrogen-fixing clade; thus, the ability to fix nitrogen may be plesiomorphic and subsequently lost in most descendants of the original nitrogen-fixing plant; however, it may be that the basic genetic and physiological requirements were present in an incipient state in the most recent common ancestors of all these plants, but only evolved to full function in some of them.

Endosymbiosis in Diatoms

Rhopalodia gibba, a diatom alga, is a eukaryote with cyanobacterial N_2-fixing endosymbiont organelles. The spheroid bodies reside in the cytoplasm of the diatoms and are inseparable from their hosts.

Nitrification

Nitrification is the biological oxidation of ammonia to nitrite followed by the oxidation of the nitrite to nitrate. The transformation of ammonia to nitrite is usually the rate limiting step of nitrification. Nitrification is an important step in the nitrogen cycle in soil. Nitrification is an aerobic process performed by small groups of autotrophic bacteria and archaea. This process was discovered by the Russian microbiologist Sergei Winogradsky.

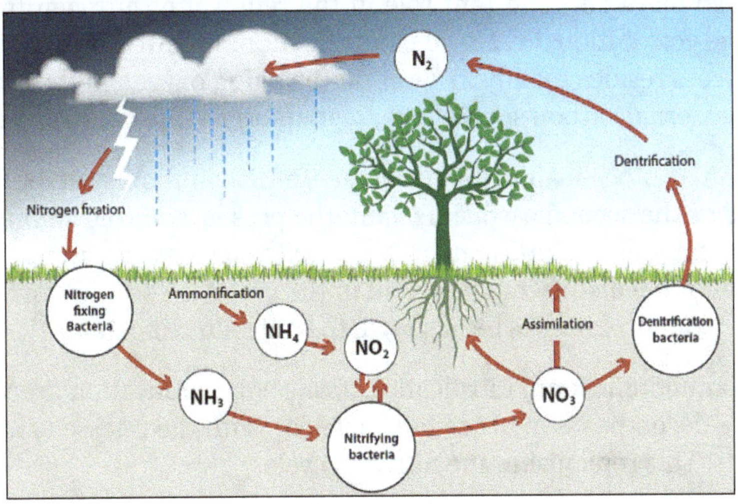

Nitrogen cycle.

Microbiology and Ecology

The oxidation of ammonia into nitrite is performed by two groups of organisms, ammonia-oxidizing bacteria (AOB) and ammonia-oxidizing archaea (AOA). AOB can be found among the β-proteobacteria and gammaproteobacteria. Currently, two AOA,

Nitrosopumilus maritimus and *Nitrososphaera viennensis*, have been isolated and described. In soils the most studied AOB belong to the genera *Nitrosomonas* and *Nitrosococcus*. Although in soils ammonia oxidation occurs by both AOB and AOA, AOA dominate in both soils and marine environments, suggesting that *Thaumarchaeota* may be greater contributors to ammonia oxidation in these environments.

The second step (oxidation of nitrite into nitrate) is done (mainly) by bacteria of the genus *Nitrobacter* and *Nitrospira*. Both steps are producing energy to be coupled to ATP synthesis. Nitrifying organisms are chemoautotrophs, and use carbon dioxide as their carbon source for growth. Some AOB possess the enzyme, urease, which catalyzes the conversion of the urea molecule to two ammonia molecules and one carbon dioxide molecule. *Nitrosomonas europaea*, as well as populations of soil-dwelling AOB, have been shown to assimilate the carbon dioxide released by the reaction to make biomass via the Calvin Cycle, and harvest energy by oxidizing ammonia (the other product of urease) to nitrite. This feature may explain enhanced growth of AOB in the presence of urea in acidic environments.

In most environments, organisms are present that will complete both steps of the process, yielding nitrate as the final product. However, it is possible to design systems in which nitrite is formed (the *Sharon process*).

Nitrification is important in agricultural systems, where fertilizer is often applied as ammonia. Conversion of this ammonia to nitrate increases nitrogen leaching because nitrate is more water-soluble than ammonia.

Nitrification also plays an important role in the removal of nitrogen from municipal wastewater. The conventional removal is nitrification, followed by denitrification. The cost of this process resides mainly in aeration (bringing oxygen in the reactor) and the addition of an external carbon source (e.g., methanol) for the denitrification.

Nitrification can also occur in drinking water. In distribution systems where chloramines are used as the secondary disinfectant, the presence of free ammonia can act as a substrate for ammonia-oxidizing microorganisms. The associated reactions can lead to the depletion of the disinfectant residual in the system. The addition of chlorite ion to chloramine-treated water has been shown to control nitrification.

Together with ammonification, nitrification forms a mineralization process that refers to the complete decomposition of organic material, with the release of available nitrogen compounds. This replenishes the nitrogen cycle.

Chemistry and Enzymology

Nitrification is a process of nitrogen compound oxidation (effectively, loss of electrons from the nitrogen atom to the oxygen atoms), and is catalyzed step-wise by a series of enzymes.

$$2NH_4^+ + 3O_2 -> 2NO_2^- + 4H^+ + 2H_2O (Nitrosomonas, Comammox)$$

$$2NO_2^- + O_2^- > 2NO_3^- \ (Nitrobacter, Nitrospira, Comammox)$$

or,

$$NH_3 + O_2^- > NO_2^- + 3H^+ + 2e^-$$

$$NO_2^- + H_2O -> NO_3^- + 2H + +2e^-$$

In *Nitrosomonas europaea*, the first step of oxidation (ammonia to hydroxylamine) is carried out by the enzyme ammonia monooxygenase (AMO).

$$NH_3 + O_2 + 2H^+ -> NH_2OH + H_2O$$

The second step (hydroxylamine to nitrite) is carried out step-wise by two different enzymes. Hydroxylamine oxidoreductase (HAO), converts hydroxylamine to nitric oxide.

$$NH_2OH -> NO + 3H^+ + 3e^-$$

Another as-of-yet unknown enzyme that converts nitric oxide to nitrite.

The third step (nitrite to nitrate) is completed in a different organism.

$$nitrite + acceptor \rightleftharpoons nitrate + reduced\ acceptor$$

Nitrification in the Marine Environment

In the marine environment, nitrogen is often the limiting nutrient, so the nitrogen cycle in the ocean is of particular interest. The nitrification step of the cycle is of particular interest in the ocean because it creates nitrate, the primary form of nitrogen responsible for "new" production. Furthermore, as the ocean becomes enriched in anthropogenic CO_2, the resulting decrease in pH could lead to decreasing rates of nitrification. Nitrification could potentially become a "bottleneck" in the nitrogen cycle.

Nitrification, as stated above, is formally a two-step process; in the first step ammonia is oxidized to nitrite, and in the second step nitrite is oxidized to nitrate. Different microbes are responsible for each step in the marine environment. Several groups of ammonia-oxidizing bacteria (AOB) are known in the marine environment, including *Nitrosomonas*, *Nitrospira*, and *Nitrosococcus*. All contain the functional gene ammonia monooxygenase (AMO) which, as its name implies, is responsible for the oxidation of ammonia. More recent metagenomic studies have revealed that some Thaumarchaeota (formerly Crenarchaeota) possess AMO. Thaumarchaeotes are abundant in the ocean and some species have a 200 times greater affinity for ammonia than AOB, leading researchers to challenge the previous belief that AOB are primarily responsible for nitrification in the ocean. Furthermore, though nitrification is classically thought to be vertically separated from primary production because the oxidation of nitrogen by bacteria is inhibited by light, nitrification by AOA does not appear to be light inhibited, meaning that nitrification is occurring throughout the water column, challenging the classical definitions of "new" and "recycled" production.

In the second step, nitrite is oxidized to nitrate. In the oceans, this step is not as well understood as the first, but the bacteria *Nitrospina* and *Nitrobacter* are known to carry out this step in the sea.

Soil Conditions controlling Nitrification Rates:

- Substrate availability (presence of NH_4^+).
- Aeration (availability of O_2).
- Well-drained soils with 60% soil moisture.
- pH (near neutral).
- Temperature (best 20-30 °C) => Nitrification is seasonal, affected by land use practices.

Inhibitors of Nitrification

Nitrification inhibitors are chemical compounds that slow the nitrification of ammonia, ammonium-containing, or urea-containing fertilizers, which are applied to soil as fertilizers. These inhibitors can help reduce losses of nitrogen in soil that would otherwise be used by crops. Nitrification inhibitors are used widely, being added to approximately 50% of the fall-applied anhydrous ammonia in states in the U.S., like Illinois. They are usually effective in increasing recovery of nitrogen fertilizer in row crops, but the level of effectiveness depends on external conditions and their benefits are most likely to be seen at less than optimal nitrogen rates.

The environmental concerns of nitrification also contribute to interest in the use of nitrification inhibitors: The primary product, nitrate, leaches into groundwater, producing acute toxicity in multiple species of wildlife and contributing to the eutrophication of standing water. Some inhibitors of nitrification also inhibit the production of methane, a greenhouse gas.

The inhibition of the nitrification process is primarily facilitated by the selection and inhibition/destruction of the bacteria that oxidize ammonia compounds. A multitude of compounds that inhibit nitrification, which can be divided into the following areas: the active site of ammonia monooxygenase (AMO), mechanistic inhibitors, and the process of N-heterocyclic compounds. The process for the latter of the three is not yet widely understood, but is prominent. The presence of AMO has been confirmed on many substrates that are nitrogen inhibitors such as dicyandiamide, ammonium thiosulfate, and nitrapyrin.

The conversion of ammonia to hydroxylamine is the first step in nitrification, where AH_2 represents a range of potential electron donors.

$$NH_3 + AH_2 + O_2 \rightarrow NH_2OH + A + H_2O$$

This reaction is catalyzed by AMO. Inhibitors of this reaction bind to the active site on AMO and prevent or delay the process. The process of oxidation of ammonia by AMO is regarded with importance due to the fact that other processes require the co-oxidation of NH_3 for a supply of reducing equivalents. This is usually supplied by the compound hydroxylamine oxidoreductase (HAO) which catalyzes the reaction:

$$NH_2OH + H_2O \rightarrow NO_2^- + 5\,H^+ + 4\,e^-$$

The mechanism of inhibition is complicated by this requirement. Kinetic analysis of the inhibition of NH_3 oxidation has shown that the substrates of AMO have shown kinetics ranging from competitive to noncompetitive. The binding and oxidation can occur on two different locations on AMO: in competitive substrates, binding and oxidation occurs at the NH_3 site, while in noncompetitive substrates it occurs at another site.

Mechanism based inhibitors can be defined as compounds that interrupt the normal reaction catalyzed by an enzyme. This method occurs by the inactivation of the enzyme via covalent modification of the product, which ultimately inhibits nitrification. Through the process, AMO is deactivated and one or more proteins is covalently bonded to the final product. This is found to be most prominent in a broad range of sulfur or acetylenic compounds.

Sulfur-containing compounds, including ammonium thiosulfate (a popular inhibitor) are found to operate by producing volatile compounds with strong inhibitory effects such as carbon disulfide and thiourea.

In particular, thiophosphoryl triamide has been a notable addition where it has the dual purpose of inhibiting both the production of urease and nitrification. In a study of inhibitory effects of oxidation by the bacteria Nitrosomonas europaea, the use of thioethers resulted in the oxidation of these compounds to sulfoxides, where the S atom is the primary site of oxidation by AMO. This is most strongly correlated to the field of competitive inhibition.

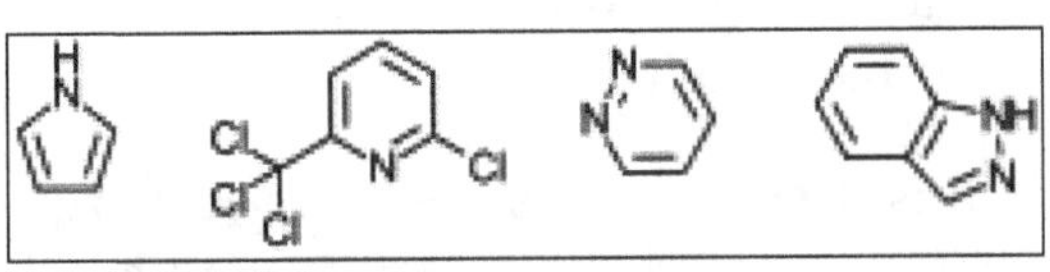

Examples of N-heterocyclic molecules.

N-heterocyclic compounds are also highly effective nitrification inhibitors and are often classified by their ring structure. The mode of action for these compounds is not well understood: while nitrapyrin, a widely used inhibitor and substrate of AMO, is a weak mechanism-based inhibitor of said enzyme, the effects of said mechanism are unable to correlate directly with the compound's ability to inhibit nitrification. It is suggested that nitrapyrin acts against the monooxygenase enzyme within the bacteria, preventing growth and CH_4/NH_4 oxidation. Compounds containing two or three adjacent ring N atoms (pyridazine, pyrazole, indazole) tend to have a significantly higher inhibition effect than compounds containing non-adjacent N atoms or singular ring N atoms (pyridine, pyrrole). This suggests that the presence of ring N atoms is directly correlated with the inhibition effect of this class of compounds.

Methane Inhibition

Some enzymatic nitrification inhibitors, such as urease, can also inhibit the production of methane in methanotrophic bacteria. AMO shows similar kinetic turnover rates to methane monooxygenase (MMO) found in methanotrophs, indicating that MMO is a similar catalyst to AMO for the purpose of methane oxidation. Furthermore, methanotrophic bacteria share many similarities to NH_3 oxidizers such as Nitrosomonas. The inhibitor profile of particulate forms of MMO (pMMO) shows similarity to the profile of AMO, leading to similarity in properties between MMO in methanotrophs and AMO in autotrophs.

Environmental Concerns

Nitrification inhibitors are also of interest from an environmental standpoint because of the production of nitrates and nitrous oxide from the process of nitrification. Nitrous oxide (N_2O), although its atmospheric concentration is much lower than that of CO_2, has a global warming potential of about 300 times greater than carbon dioxide and contributes 6% of planetary warming due to greenhouse gases. This compound is also notable for catalyzing the breakup of ozone in the stratosphere. Nitrates, a toxic compound for wildlife and livestock and a product of nitrification, are also of concern.

Soil, consisting of polyanionic clays and silicates, generally has a net anionic charge. Consequently, ammonium (NH_4^+) binds tightly to the soil but nitrate ions (NO_3^-) do not. Because nitrate is more mobile, it leaches into groundwater supplies through agricultural runoff. Nitrates in groundwater can affect surface water concentrations, either through direct groundwater-surface water interactions (e.g., gaining stream reaches, springs), or from when it is extracted for surface use. As an example, much of the drinking water in the United States comes from groundwater, but most wastewater treatment plants discharge to surface water.

Wildlife such as amphibians, freshwater fish, and insects are sensitive to nitrate levels, and have been known to cause death and developmental anomalies in affected species. Nitrate levels also contribute to eutrophication, a process in which large algal blooms reduce oxygen levels in bodies of water and lead to death in oxygen-consuming creatures due to anoxia. Nitrification is also thought to contribute to the formation of photochemical smog, ground level ozone, acid rain, changes in species diversity, and other undesirable processes. In addition, nitrification inhibitors have also been shown to suppress the oxidation of methane (CH_4), a potent greenhouse gas, to CO_2. Both nitrapyrin and acetylene are shown to be especially strong suppressors of both processes, although the modes of action distinguishing them are unclear.

Denitrification

Denitrification is a microbially facilitated process where nitrate (NO_3^-) is reduced and ultimately produces molecular nitrogen (N_2) through a series of intermediate gaseous nitrogen oxide products. Facultative anaerobic bacteria perform denitrification as a type of respiration that reduces oxidized forms of nitrogen in response to the oxidation of an electron donor such as organic matter. The preferred nitrogen electron acceptors in order of most to least thermodynamically favorable include nitrate (NO_3^-), nitrite (NO_2^-), nitric oxide (NO), nitrous oxide (N_2O) finally resulting in the production of dinitrogen (N_2) completing the nitrogen cycle. Denitrifying microbes require a very low oxygen concentration of less than 10%, as well as organic C for energy. Since denitrification can remove NO_3^-, reducing its leaching to groundwater, it can be strategically used to treat sewage or animal residues of high nitrogen content. Denitrification can leak N_2O, which is an ozone-depleting substance and a greenhouse gas that can have a considerable influence on global warming.

The process is performed primarily by heterotrophic bacteria (such as *Paracoccus denitrificans* and various pseudomonads), although autotrophic denitrifiers have also been identified (e.g., *Thiobacillus denitrificans*). Denitrifiers are represented in all main phylogenetic groups. Generally several species of bacteria are involved in the complete reduction of nitrate to N_2, and more than one enzymatic pathway has been identified in the reduction process.

Direct reduction from nitrate to ammonium, a process known as dissimilatory nitrate reduction to ammonium or DNRA, is also possible for organisms that have the nrf-gene. This is less common than denitrification in most ecosystems as a means of nitrate reduction. Other genes known in microorganisms which denitrify include *nir* (nitrite reductase) and *nos* (nitrous oxide reductase) among others; organisms identified as having these genes include *Alcaligenes faecalis*, *Alcaligenes xylosoxidans*, many in the genus *Pseudomonas*, *Bradyrhizobium japonicum*, and *Blastobacter denitrificans*.

Half Reactions

Denitrification generally proceeds through some combination of the following half reactions, with the enzyme catalyzing the reaction in parentheses:

- $NO_3^- + 2\,H^+ + 2\,e^- \rightarrow NO_2^- + H_2O$ (Nitrate reductase).

- $NO_2^- + 2\,H^+ + e^- \rightarrow NO + H_2O$ (Nitrite reductase).
- $2\,NO + 2\,H^+ + 2\,e^- \rightarrow N_2O + H_2O$ (Nitric oxide reductase).
- $N_2O + 2\,H^+ + 2\,e^- \rightarrow N_2 + H_2O$ (Nitrous oxide reductase).

The complete process can be expressed as a net balanced redox reaction, where nitrate (NO_3^-) gets fully reduced to dinitrogen (N_2):

- $2\,NO_3^- + 10\,e^- + 12\,H^+ \rightarrow N_2 + 6\,H_2O$.

Conditions of Denitrification

In nature, denitrification can take place in both terrestrial and marine ecosystems. Typically, denitrification occurs in anoxic environments, where the concentration of dissolved and freely available oxygen is depleted. In these areas, nitrate (NO_3^-) or nitrite (NO_2^-) can be used as a substitute terminal electron acceptor instead of oxygen (O_2), a more energetically favourable electron acceptor. Terminal electron acceptor is a compound that gets reduced in the reaction by receiving electrons. Examples of anoxic environments can include soils, groundwater, wetlands, oil reservoirs, poorly ventilated corners of the ocean and seafloor sediments.

Furthermore, denitrification can occur in oxic environments as well. High activity of denitrifiers can be observed in the intertidal zones, where the tidal cycles cause fluctuations of oxygen concentration in sandy coastal sediments. For example, the bacterial species *Paracoccus denitrificans* engages in denitrification under both oxic and anoxic conditions simultaneously. Upon oxygen exposure, the bacteria is able to utilize nitrous oxide reductase, an enzyme that catalyzes the last step of denitrification. Aerobic denitrifiers are mainly Gram-negative bacteria in the phylum Proteobacteria. Enzymes NapAB, NirS, NirK and NosZ are located in the periplasm, a wide space bordered by the cytoplasmic and the outer membrane in Gram-negative bacteria.

Denitrification can lead to a condition called isotopic fractionation in the soil environment. The two stable isotopes of nitrogen, ^{14}N and ^{15}N are both found in the sediment profiles. The lighter isotope of nitrogen, ^{14}N, is preferred during denitrification, leaving the heavier nitrogen isotope, ^{15}N, in the residual matter. This selectivity leads to the enrichment of ^{14}N in the biomass compared to ^{15}N. Moreover, the relative abundance of ^{14}N can be analyzed to distinguish denitrification apart from other processes in nature.

Use in Wastewater Treatment

Denitrification is commonly used to remove nitrogen from sewage and municipal wastewater. It is also an instrumental process in constructed wetlands and riparian zones for the prevention of groundwater pollution with nitrate resulting from excessive agricultural or residential fertilizer usage. Wood chip bioreactors have been studied since the 2000s and are effective in removing nitrate from agricultural run off and even manure.

Reduction under anoxic conditions can also occur through process called anaerobic ammonium oxidation (anammox):

$$NH_4^+ + NO_2^- \rightarrow N_2 + 2\ H_2O$$

In some wastewater treatment plants, compounds such as methanol, ethanol, acetate, glycerin, or proprietary products are added to the wastewater to provide a carbon and electron source for denitrifying bacteria. The microbial ecology of such engineered denitrification processes is determined by the nature of the electron donor and the process operating conditions. Denitrification processes are also used in the treatment of industrial wastewater. Many denitrifying bioreactor types and designs are available commercially for the industrial applications, including Electro-Biochemical Reactors (EBRs), membrane bioreactors (MBRs), and moving bed bioreactors (MBBRs).

Aerobic denitrification, conducted by aerobic denitrifiers, may offer the potential to eliminate the need for separate tanks and reduce sludge yield. There are less stringent alkalinity requirements because alkalinity generated during denitrification can partly compensate for the alkalinity consumption in nitrification.

Commercialization of Agriculturally Important Microorganisms

Microbes of the phytomicrobiome are associated with every plant tissue and, in combination with the plant form the holobiont. Plants regulate the composition and activity of their associated bacterial community carefully. These microbes provide a wide range of services and benefits to the plant; in return, the plant provides the microbial community with reduced carbon and other metabolites. Soils are generally a moist environment, rich in reduced carbon which supports extensive soil microbial communities. The rhizomicrobiome is of great importance to agriculture owing to the rich diversity of root exudates and plant cell debris that attract diverse and unique patterns of microbial colonization. Microbes of the rhizomicrobiome play key roles in nutrient acquisition and assimilation, improved soil texture, secreting, and modulating extracellular molecules such as hormones, secondary metabolites, antibiotics, and various signal compounds, all leading to enhancement of plant growth. The microbes and compounds they secrete constitute valuable biostimulants and play pivotal roles in modulating plant stress responses. Research has demonstrated that inoculating plants with plant-growth promoting rhizobacteria (PGPR) or treating plants with microbe-to-plant signal compounds can be an effective strategy to stimulate crop growth. Furthermore, these strategies can improve crop tolerance for the abiotic stresses (e.g., drought, heat, and salinity) likely to become more frequent as climate change conditions continue to develop. This discovery has resulted in multifunctional PGPR-based formulations for commercial agriculture, to minimize the use of synthetic fertilizers and agrochemicals.

A plant growing under field conditions is not an individual; it is a complex community with subtle and relatively constant partner relationships. A well-structured and regulated community of microorganisms is always associated with the plant. This community is the phytomicrobiome; the phytomicrobiome plus the plant is the holobiont. Microbiome relationships exist with all multi-cellular organisms, and probably all eukaryotes. In fact, these probably predate the colonization of the land by plants. This microbial community has been associated with terrestrial plants since their earliest evolution, to assist early land plants faced with challenges such as access to nutrients, novel and often-stressful conditions and pathogens.

There are elements (including bacteria and fungi) of the phytomicrobiome associated will all major plant structures (flowers, fruits, stems, leaves, and roots). However, conditions vary substantially among these structures, leading to specialized microbial populations inhabiting each one. The microbial community associated with the roots (the rhizomicrobiome), is the most populous and elaborate of all those associated with higher plants. The best understood and characterized example is the nitrogen-fixing rhizobia associated with legumes. Many members of the phytomicrobiome cannot be cultured and it has only been since the advent of metagenomics and related methods that we are able to assess how membership is changed by conditions, plant genotype and plant development.

The plant exerts considerable control over the composition of the rhizomicrobiome. It produces root exudates of various compositions, which can be more suitable as a source of reduced C, to some microbes than others. The plant also produces signal compounds that recruit specific species and regulate their genetic and biochemical activities. In addition, the soil microbial community undertakes various aspects of self-regulation. The microbes can produce quorum sensing compounds to communicate when conditions warrant a collective physiological shift. Plants have evolved to respond to microbial quorum sensing compounds and to produce analogs, providing plants with another level of regulation over the rhizomicrobiome. Finally, it is now becoming apparent that there is some degree of hierarchy within the phytomicrobiome and that there are key members, termed "hub species" or "core species", whose activities are regulated by plants, and hub species in turn regulate broader activities within the phytomicrobiome. Most hub species have probably been part of the phytomicrobiome for a very long time, allowing for development of their central position.

In the soil, there is a gradient of intimacy between plant roots and microbes extending away from the plant root: The degree of plant influence over the microbial community increases nearer the root surface. This zone is now generally referred to as the rhizosphere, however, the term was originally coined by Hiltner to describe the soil microorganisms around and inside roots. Now, microbes living on the root surface are said to inhabit the rhizoplane, and those living inside the root are said to be endophytes. Mitochondria and plastids (including the chloroplasts) represent some of the oldest and most intimate, aspects of the phytomicrobiome. They evolved from plant-associated microbes into the permanent subcellular structures we see today.

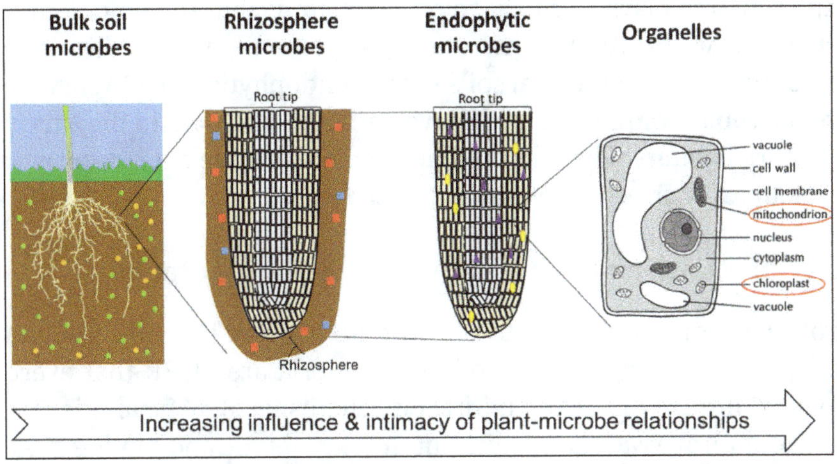

The degree of intimacy and influence of the plant-microbe interactions. Microbes are represented by small colored (red, green, yellow, purple, and blue) shapes. Diversity and number of microbes is variable between soils, distance from plant roots, crop species, and plant tissue.

Agricultural Context: A "Fresh" Green Revolution in the Face of Climate Change

The Green Revolution of the 20th century enabled unprecedented gains in global food production. The Green Revolution was roughly comprised of two main advances; chemical inputs (pesticides, herbicides, and chemical fertilizers) and improved crop plants (through targeted breeding and advanced genetic manipulations). However, gains associated with fertilizer inputs carry high environmental costs. A new revolution in agricultural innovation will be needed to sustain the food, fiber, and fuel needs of a growing global population and a changing climate through the 21st century. A "Fresh" Green Revolution, perhaps the Bio-Revolution, needs to be based on fewer intensive inputs with reduced environmental impact. A Bio-Revolution could be based on 1) biological inputs through utilization of the phytomicrobiome (with inoculants, microbially produced compounds, etc.), and improved crops (by manipulation of the phytomicrobiome community structure). The use of microbial based agricultural inputs has a long history, beginning with broad-scale rhizobial inoculation of legumes in the early 20th century. More recently, strains of Bacillus, Pseudomonas, Glomus, and others have been commercialized. The use of bacterial taxa in plant production has been reviewed previously for Bacillus, Pseudomonas, Actinobacteria, and Lactobacillus. In addition, Acetobacter, Azospirillum, Paenibacillus, Serratia, Burkholderia, Herbaspirillum, and Rhodococcus have also been shown to enhance crop production.

The effects of climate change are expected to impose more environmental stresses on crops worldwide. Moreover, as climate change progresses throughout the 21st century, significant areas of high-quality agricultural lands will likely be lost to rising seas, erosion, salinization, and desertification. This means that crop yields will need to be maintained, in spite of production on a smaller area of land, under more stressful conditions. The phytomicrobiome plays a critical role in the survival of the holobiont, particularly for plants growing in extreme environments. Some plants that live in hypersaline coastal environments or geothermal soils rely on endophytic fungi to survive. Likewise, constitutive microbial communities of agave and cacti likely aid in the survival of these plants in very dry habitats. The microbiomes of plants native to extreme environments may be rich sources of stress-ameliorating microbes.

Physiology of Plant-growth Promoting Rhizobacteria

Plant-microbe co-evolution has led to some of the bacteria becoming facultative intracellular endophytes. Among these free-living bacteria are PGPR that exert beneficial effects on plants through direct and indirect mechanisms. Beneficial rhizobacteria have been utilized to improve water and nutrient uptake, abiotic and biotic stress tolerance. Even though numerous soil bacteria have been reported to promote plant growth and development, the modes of action by which the bacteria exhibit beneficial activities are often not well understood. The molecular basis of plant-bacteria interaction mechanisms responsible for the physiological changes are beginning to be discerned, mainly due to the emerging "omics" approaches.

Nutrient Acquisition by PGPR

Soils with dynamic microbial ecologies and high organic matter typically have lower fertilizer requirements than conventionally managed soils. For example, bulk microbial activity in soils is often considered when managing the application of organic nutrient sources. Phytomicrobiome research is beginning to reveal specific plant-microbe interactions that directly aid in plant nutrition. Microbes that assist in plant nutrient acquisition (biofertilizers) act through a variety of mechanisms including augmenting surface area accessed by plant roots, nitrogen fixation, P-solubilization, siderophore production and HCN production. Therefore, manipulating microbial activity has great potential to provide crops with nutritional requirements.

The most extensively studied and exploited beneficial plant-bacteria relationship is the N-fixing symbiosis between rhizobia and legumes. In this relationship, legumes provide rhizobia with reduced C and a protected, anaerobic environment required for nitrogenase activity, while rhizobia provide the legumes with biologically available N. Within this symbiosis, both rhizobia and legume undergo significant transformations. The legume forms a new organ, the nodule, to house the rhizobia, and the rhizobia, in turn, changes from its free-living rod-shaped cell type to a branched, N-fixing bacteroid. Rhizobial N-fixation contributes significant amounts of N to global agricultural systems, with estimates ranging from 20 to 22 Tg N per year up to 40 Tg N per year. Rhizobial inoculants of leguminous crops are the earliest example of commercial microbial products in agriculture and still represent the most widely used agricultural inoculants. However, genetic improvements in efficiency of the N-fixing symbiosis of rhizobia and crop plants have been elusive. The fixation of atmospheric nitrogen and conversion to ammonia is an energy demanding process, which means oxidative phosphorylation of carbon sources to generate ATP must be favored over glycogen synthesis within the bacterial cell, to increase nitrogen fixation. However, experiments with glycogen synthase deletion mutants of Rhizobium tropici have not survived in soil environments, despite increased dry matter and nodule number in inoculated bean plants.

Starting in the early 21st century, interest began to mount around the development of commercial inoculants of free-living N-fixing bacteria such as Azoarcus sp., Burkholderia sp., Gluconacetobacter sp., Diazotrophicus sp., Herbaspirillum sp., Azotobacter sp., Bacillus polymyxa, and especially Azospirillum sp. These free-living diazotrophs provide N to a much wider range of crop plants than rhizobia. Commercial inoculants of Azospirillum, produced by small and medium sized companies around the world, have been effective in increasing the yield of various cereal crops. Other bacteria that do not fix N have been shown to increase N uptake in plants, thus increasing nitrogen use efficiency, likely due to increased root growth, which allows plants to access more soil.

Following Liebig's law of the minimum in mind, the next most limiting nutrient for crop plants after N is usually P. While most agricultural soils contain ample quantities of P, much of it is in non-soluble forms. To supplement indigenous soil P, crops are typically fertilized with rock phosphate mined from one of a few large deposits (up to 85% of the world's rock phosphate is estimated to be in Morocco and Western Sahara). Furthermore, phosphorus solubilizing microorganisms (PSMs) can help plants access the reservoir of non-labile phosphorus by releasing it from its recalcitrant forms. Inorganic P complexed with Ca, Fe, or Al can be solubilized by organic acids or H+ ions excreted by PSMs. Similarly, phytase produced by PSMs can liberate reactive P from organic compounds. Production of HCN by PGPR was originally thought to promote plant growth by suppressing pathogens, however, this idea has recently been challenged by Rijavec and Lapanje, who argued that HCN indirectly increases P availability by metal chelation and sequestration of these geochemical entities. PSMs produce organic acids to reduce metal toxicity by using these compounds to transform metal species to immobile forms or chelate them for mobility, to be carried into the plant tissues for further phyto-extraction possibilities. The PSM Bacillus megaterium has been commercialized as BioPhos and can reduce phosphate fertilizer requirements of plantation crops up to 75%. Strains of P- solublizing Pseudomonas striata, B. Polymyxa, and B. megaterium have also been commercialized by AgriLife.

Other nutrient elements, such as Fe and Zn can limit crop yields. Like P, Fe can also be abundant in soils, but unavailable to plants. Many bacterial strains increase the availability of Fe through the production of organic acids or siderophores. Siderophores also act to control pathogenic microbes by depriving them Fe. A commercial formulation of the Fe mobilizing bacteria, Acidithiobacillus ferrooxidans has been developed by AgriLife, although this genus apparently solubilizes Fe through organic acid production rather than with siderophores. Several strains of Zn-mobilizing bacteria have been shown to increase Zn uptake, and thus increase yield in several crops, including rice, wheat and soybean. While the mechanisms of Zn-mobilizers remain uncertain, they are likely similar to those of PSMs and Fe-mobilizers, namely the production of chelating agents and organic acids.

Signal Exchange between Plant Roots and PGPR

Plant Hormones Produced by PGPR

Phytohormones are key players in regulating plant growth and development. They also function as molecular signals in response to environmental factors that otherwise limit plant growth or become lethal when uncontrolled. Many rhizosphere bacteria are known to excrete hormones for root uptake or manipulate hormone balance in the plants to boost growth and stress response.

Many PGPR can produce auxins to exert particularly strong effects on root growth and architecture. Indole-3-acetic acid (IAA) is the most widely studied auxin produced by PGPR. It is involved in plant-microbe interactions. The function of exogenous IAA is dependent on the endogenous IAA levels in plants. At optimal IAA concentrations in plants, application of bacterial IAA may have neutral, positive, or negative effects on plant growth. PGPR that produce auxins have been shown to elicit transcriptional changes in hormone, defense-related, and cell wall related genes, induce longer roots, increase root biomass and decrease stomata size and density and activate auxin response genes that enhance plant growth.

Many PGPR produce cytokinins and gibberellins but the role of bacterially synthesized hormones in plants, and bacterial mechanism of synthesis, are not yet completely understood. Some strains of PGPR can promote relatively large amounts of gibberellins, leading to enhanced plant shoot growth. Interactions of these hormones with auxins can alter root architecture. Production of cytokinins by PGPR can also lead to enhanced root exudate production by the plant potentially increasing the presence of PGPR associated with the plant.

Ethylene is a gaseous hormone, active at extremely low concentrations (0.05 mL L-1) and is a “stress hormone,” as illustrated by its concentration spiking during various abiotic and biotic stresses. Accumulation of ethylene in response to stress may increase plant tolerance or exacerbate stress-response symptoms and senescence. PGPR function has been studied under both stressed and unstressed conditions and often provides greater growth stimulation under stressful conditions, for instance, under drought stress. Ethylene plays an important role for improving plant stress tolerance for some PGPR: PGPR secrete 1-aminocyclopropane-1-carboxylase (ACC) deaminase which reduces ethylene production in plants. Many studies have shown enhanced stress tolerance in plants through inoculation with PGPR that produce ACC deaminase. This appears to occur since PGPR are able to keep ethylene levels from reaching levels sufficient to reduce plant growth, as has been demonstrated with Camelina sativa.

Other Microbe-to-plant Signal Molecules

A wide range of secondary metabolites and volatile organic compounds (VOCs) produced by bacteria can improve stress-tolerance and stimulate growth in plants. For example, polyamines play important physiological and protective roles in plants. B. megaterium $BOFC_{15}$ secretes a polyamine, spermidine, and induces polyamine production in Arabidopsis, resulting in an increase in biomass, altered root architecture and elevated photosynthetic capacity. The inoculated plants exhibited higher drought tolerance and abscisic acid (ABA) content under PEG induced water-deficit stress. A range of PGPR produce HCN, which can control the level of deleterious microbes in the rhizosphere. VOC produced by PGPR stimulate plant growth, resulting in increased shoot biomass and improve plant stress resistance.

The microbes of the phytomicrobiome also affect each other's activities through signal compounds. These signals amount to hormones of the holobiont. For example, lumichrome and riboflavin can act as microbe-to-plant signal compounds able to stimulate plant growth. Both compounds can cause meaningful alterations in plant development; lumichrome can accelerate appearance of leaves (more rapid development) and leaf expansion (enhanced growth). In addition, it can increase plant height and overall leaf area, resulting in improved production of biomass. This is true over a wide range of plant types including both monocots and dicots.

Microbe-to-plant signal compounds (e.g., lipo-chitooligosaccharides and thuricin 17) have been shown to increase plant growth for diverse species, particularly when plants are growing under stressful conditions. The receptor for the lipo-chitooligosaccharides is a LysM kinase for the legume-rhizobia symbioses; this receptor system seems to have evolved for pathogen detection almost Two billion years ago. The microbe-to-plant signal in the N_2-fixing Frankia symbiosis remains to be identified but appears not to be an LCO.

Root Exudates as Plant-to-microbe Signals

Plants excrete considerable control over the microbes they associated with; even of simple genotype differences within a plant species can have meaningful effects. Some of this control is the result of inter-organismal signals. Starting when the seed is imbibing and germinating, then when roots are growing and finally senescing, molecules are released from roots into the surrounding soil. These molecules support microbial growth and activity in the rhizosphere. Variation in root exudation (timing, amount, and constituents) provides a mechanism by which plants can manipulate composition and abundances of their root-associated microbiota. Exudates are thought to consist mainly of sugars, amino acids, and organic acids that are present at high concentrations in the cytoplasm of the plant, but also include smaller amounts of complex secondary metabolites such as flavonoids, terpenes, and phenolic compounds that can attract specific microbes in the rhizosphere. It has also been suggested that exudation of the signal molecules jasmonic acid and salicylic acid into the rhizosphere can be involved in the interplay between roots and microbes during the initial events of colonization. Root exudation is genetically regulated and can thus shape distinct rhizobacterial communities for different plant genotypes, resulting in highly variable exudates among plant species, individual plant types within the same species, at different plant developmental stages, growth conditions, and biotic interactions.

PGPR Improve Plant Growth under Stressful Growing Conditions

The mechanisms that regulate stress tolerance in plants are intricate and complex, in part because plants are sessile organisms which have no choice but to stand where they are and "take it." Improving stress tolerance in crop plants through conventional breeding is a long and capital-intensive process, while genetic engineering is associated with ethical and social acceptance issues. The role of beneficial microorganisms is gaining importance in stress management and the development of climate change resilient agriculture. Recent studies have exploited molecular techniques to understand the mode of action of the plant-microbe interactions resulting in induced stress tolerance.

Abiotic Stress Tolerance Associated with PGPR

Pseudomonas putida MTCC5279 ameliorated drought stress in chickpea (Cicer arietinum) plants by modulating membrane integrity, osmolyte accumulation (proline, glycine betaine) and ROS scavenging ability. Stress responses were positively modulated by the bacteria resulting in differential expression of genes involved in ethylene biosynthesis (ACO and ACS), salicylic acid (PR1), jasmonate (MYC2) transcription activation, SOD, CAT, APX, and GST (code for antioxidant enzymes), DREB1A (dehydration responsive element binding), NAC1 (transcription factors expressed under abiotic stress), LEA and DHN (dehydrins). Application of thuricin 17 produced by Bacillus thuringiensis NEB17 to soybean (Glycine max) under water-deficit conditions resulted in modification of root structures and increased root and nodule biomass, root length, root ABA, and total nitrogen content. Beneficial microbes also help plants cope with flooding stress. Rice (Oryza sativa) seedlings inoculated with an ACC deaminase producing strain of Pseudomonas fluorescens REN1 increased root elongation under constantly flooded conditions.

Salt stress effects can be diminished by ACC deaminase. Pea plants inoculated with Variovorax paradoxus 5C-2, which produce ACC deaminase, had increased photosynthetic rate, electron transport, balanced ion homeostasis through increased $K+$ flow to shoots and $Na+$ deposition on roots, decreased stomatal resistance and xylem balance pressure and increased biomass under salt stress at 70 and 130 mM NaCl. For okra, PGPR producing ACC enhanced salt tolerance, increased antioxidant enzyme activities (SOD, APX, and CAT) and upregulated ROS pathway genes (CAT, APX, GR, and DHAR). Maize seedlings inoculated with Bacillus amyloliquefaciens SQR9, had enhanced salt stress tolerance, including enhanced the chlorophyll content, compared with the control. Additional analysis showed that the mechanisms were related to enhanced total soluble sugar content leading to decreased cell destruction, improved peroxidase/catalase activity and glutathione content for scavenging ROS, and reduced Na+ levels in the plant. These physiological manifestations were confirmed by measured upregulation of RBCS, RBCL, H+ -PPase, HKT1, NHX1, NHX2, and NHX3 genes, as well as downregulation of NCED expression, as determined by qPCR. Wheat (Triticum aestivum) plants inoculated with the halotolerant Dietzia natronolimnaea showed upregulation of genes involved in the ABA-signaling cascade, salt overly sensitive (SOS) pathway, ion transporters, and antioxidant enzymes; stress tolerance is induced by modulation of complex network of gene families.

Exposure to cold and heat reduce yield and, in worst case scenarios, result in crop failure. A gibberellin-producing PGPR, Serratia nematodiphila increases pepper (Capsicum annum) growth under low temperature stress conditions. The inoculated plants contained more GA4 and ABA and less salicylate and jasmonate. Inoculation with Burkholderia phytofirmans PsJN modulated carbohydrate metabolism to reduce chilling damage to grapevine (Vitis vinifera) plantlets exposed to low temperature stress. Inoculation of tomato (Solanum lycopersicum) plants exposed to low temperatures with Pseudomonas vancouverensis OB155 and P. frederiksbergensis OS261 increased expression of cold acclimation genes and antioxidant activity in leaf tissues.

Biocontrol and Induced Systemic Resistance for Biotic Stress Tolerance

Bacillus amyloliquefaciens (SN13) is a biocontrol agent against Rhizoctonia solani, by prolonging tolerance through enhanced defense response in the plants. The colonized plants exhibit modulation of phytohormone signaling, sustained maintenance of elicitors, production of secondary metabolites and balance of reactive oxygen species and scavengers producing ROS scavengers. Cotton (Gossypium hirsutum) plants inoculated with Bacillus spp. exhibited increased gossypol and jasmonic acid secretion reducing larval feeding by Spodoptera exigua. Transcript levels of genes involved in synthesis of allelochemicals and jasmonates were higher in inoculated plants as was suppression of the pest. Enterobacter asburiae BQ9 induced resistance against tomato yellow leaf curl virus by increasing the expression of defense-related genes and antioxidant enzymes, including phenylalanine ammonia lyase, peroxidase, catalase, and superoxide dismutase. Soil inoculation with Peanibacillus lentimorbus B-30488 decreased cucumber mosaic virus RNA accumulation in Nicotiana tabacum cv. White burley leaves by 91%. This was associated with an increase in stress and pathogenesis-related gene expression and antioxidant enzyme activity suggesting induced resistance against the virus. PGPR colonization resulted in improved tissue heath and physiology of plants, which produced more flowers and seeds. The bacteria also produce ACC deaminase and induce tolerance against southern blight disease in tomato caused by Scelerotium rolfsii. The inoculated plants showed modulation of the ethylene pathway and antioxidant enzyme activities; systemic tolerance was corroborated by pathogen related gene expression analysis. Acyl-homoserine lactones (AHL)-producing Serratia liquefaciens MG1 and P. putida IsoF elicited induced systemic resistance (ISR) in tomato (S. lycopersicum) against Alternaria alternate whereas AHL-null mutant strains of both PGPR resulted in reduced ISR. Root exudates have been found to contain chemicals that mimic AHL signals, stimulating beneficial rhizosphere associations while inhibiting pathogenic bacteria.

Besides functioning as biocontrol agents, PGPR protect plants against pathogens by eliciting biochemical and molecular defense responses within the plant. PGPR can trigger ISR in plants, which activates pathogenesis-related genes, mediated by phytohormone signaling pathways and defense regulatory proteins to prime plants against future pathogen attack. Bacterial signal compounds and microbe-associated molecular triggers, such as chitin oligomers, have been shown to modulate ISR induction in plants. Pathogen cell-surface factors such as flagellins and O-antigen of lipopolysaccharides elicit ISR, whereas analogs of salicylic acid and jasmonic acid trigger ethylene to elicit NPR1 mediated systemic acquired resistance (SAR) in plants.

Strategies for Improving Rhizosphere Colonization by PGPR Inoculants

Under field conditions, other external factors come into play and the ability of soil bacteria to elicit positive effects on plant growth can be impaired and so that the effects of applying specific PGPM can be variable. The plant rhizosphere is colonized by microorganisms from the soil and the seed. The determinants of soil microorganisms are based on properties such as C and N availability, organic matter content, water availability and pH as well as biogeographic patterns including soil type and seasonality. Hence it is necessary to develop strategies for effective inoculation methods, so that bacteria of interest gain advantage in colonization efficiency over others. Product quality, compatibility, and stability determine effective colonization and consistent performance of the inoculum under field conditions.

Biofilm versus Planktonic Inoculum

Plant-associated biofilms have been shown to establish themselves on various parts of plants such as leaves, roots, seeds and internal vasculature. The ability to form biofilms not only enhances bacterial survival but also enhances plant growth through the various PGPR-associated mechanisms, often to a greater extent than their planktonic cell counterparts. Another advantage of biofilms over planktonic cells is their higher resistance to antibiotics, leading to improved chance of survival in a competitive soil environment. This is an important consideration when applying microbial inoculants to soils where microbes face intense competition and may not be as well adapted to challenging conditions as indigenous soil. An alternative mechanism by which biofilms enhance plant growth is through biocontrol of disease organisms, such as competitive colonization of the rhizosphere and the production of antimicrobial compounds.

Single and dual-species biofilms produced from Pseudomonas, Trichoderma, Bradyrhizobium, and Penicillium showed greater ammonia production, IAA production, phosphate solubilization, siderophore production, and nitrogenase activity than the planktonic inocula. Furthermore, when the biofilms were used to inoculate seeds, cotton seed germination, wheat root and shoot length, soybean dry weights and nitrogen accumulation, and maize seed germination and root length were increased compared to plants inoculated with planktonic cells.

Using Biochar to Promote Microbial Growth and Survival in Soil

Biochar has received much attention in the scientific literature over the last decade, as a soil amendment due to its ability to improve soil fertility and increase crop yields. Biochar can change soil fertility parameters that influence microbial survival in soil, including pH, organic matter content, cation exchange capacity and nutrient retention, water retention and oxygen tension, bulk density and provide niche spaces for microbes, thus preventing grazing by fungal predators. Recent research has also investigated the use of biochar as a carrier material for microbial inoculants, applied as seed-coatings, constituting a sustainable alternative to peat-based inoculants, and promoting early colonization of the rhizosphere with beneficial microorganisms. It is important to note, however, that not all biochar materials are the same; biochar production conditions and feedstock materials have a large influence on the biological, chemical and physical properties of the final biochar material and while many provide desirable effects on soil fertility, some can be toxic to microbes and plants.

Challenges Moving from the Lab to the Field

While the technology of bio-inoculants holds a promising future, some major bottle necks have to be addressed to increase their efficacy. The use of PGPRs as inoculants is centuries old; the use of these inoculants have been largely focussed on legumes and cereals. Development of new PGPR inocula is based on laboratory screening assays that rely on specific PGPR mechanisms, namely nitrogen fixation, ACC deaminase activity, auxin synthesis and calcium phosphate solubilization. However, screening of pure culture isolates for those with PGPR functions does not always result in isolates that promote plant growth under field conditions. At the same time, those which have minimal in vitro growth promoting functions may have alternate mechanisms to promote plant growth. Since these mechanisms are less well-understood, they are difficult to screen for under laboratory conditions. As a result, beneficial strains that employ these mechanisms are discarded based on poor performance on classical in vitro PGPR screening methods.

Developing inocula containing highly effective microbes with a long shelf-life and high rhizosphere colonization rate poses a major challenge for commercialization. PGPR are often used to inoculate plant material without an appropriate carrier or in quantities that do not allow for efficient rhizosphere colonization under field conditions, due to competition with resident soil micro- and macro-fauna. In addition, soils growing high value crops are often fumigated with broad spectrum biocidal fumigants that alter the bio-community structure of the soil. Long-term fumigation affects soil microbes and their interactions that help plants with nutrient acquisition and mobilization, thereby affecting soil health. This may also pose a challenge to rhizosphere colonization by PGPR inocula.

Plant breeding has been instrumental in the success of Green Revolution. However, in the context of bio-inoculants, very little has been done to integrate microbiome-based plant breeding to achieve a heritable PGPR community that enhances crop productivity. The Green Revolution also has introduced inorganic fertilizers, pesticides, and herbicides into soils leading to extensive damage in the form of contaminants. Combining bioremediation with plant growth promotion would be a beneficial approach in addressing this global agriculture problem. Designing microbial consortia to address various aspects of bioremediation and plant growth potential is an essential aspect to this approach. Synthesis of bio-inoculants for specific soil conditions, to overcome environmental constraints, and training farmers and associated staff to efficiently apply them to crop plants is very important element in the development and deployment of more beneficial inocula.

Applications

Bacteria with multiple benefits can be advantageous in commercial agriculture and are relevant to the bio-economy. Many plants of economic significance are grown in monoculture and require amendments for optimal growth and yield, as well as protection against disease organisms.

Increasing Yield and Decreasing Fertilizer Inputs

Utilization of bacterial consortia has inconsistent effects on crop yield. The mixing of a bacterium (B. amyloliquefaciens) with a fungus (Trichoderma virens) improves yields of corn and tomato, among other crops and is available in the market place. The company Excalibre-SA (ABM) combines Trichoderma with Bradyrhizobium for improved growth of soybean while BioGrow Endo (Mycorrhizal Applications) combines arbuscular mycorrhizal fungi and Trichoderma for improved growth and treatment of pathogens present in the soil; both of which are commercially available.

Inoculation with N-fixing bacteria (Azospirillum and Azobacter) allowed half-rate N-fertilizer application and increased sesame seed yield and oil quality. Similar effects were shown for Azospirillum vinelandii inoculated Brassica carinata cv. Peela raya. A consortium of bacteria (Bacillus cereus PX35, Bacillus subtilis SM21, and Serrati asp XY2) reduced the incidence of root knot nematode (Meloidogyne incognito) in tomato, increased fruit yield (31.5 to 39%) and quality (soluble sugars, vitamin C, and titratable acids.)

Advanced biofuels are derived from non-food biomass, often lignocellulosic material, to minimize any competition with food production; the long-term goal is provision of renewable fuels, along with high value bio-products, to reduce the atmospheric CO2 emissions associated with fossil fuels. Conversion of lignocellulosic material to fuel needs to become easier and less expensive to make this fuel economically competitive; in addition, there needs to be improved biomass availability from purpose-grown biomass crops (e.g., Miscanthus, switchgrass, and Sorghum bicolour). The growth and productivity of purpose grown biofuel crops can be improved through inoculation with PGPR as has been demonstrated for switchgrass. Marginal and contaminated lands can be used to grow biofuel crops in order to avoid conflicts around food versus energy crops. With the use of PGPR that contain natural potential to cope with soil contaminants, the biofuel crops could be used efficiently for phytoremediation and also to reduce high levels of agrochemicals residues in agriculture lands.

Improving Disease Control and Reducing the use of Agrochemicals

Biologicals are an alternative method for combating plant pathogens, and there are commercially available examples. Beneficial rhizobacteria may secrete antibiotics and other compounds antagonistic to plant pathogens. Production of antibiotics is one of the more common biocontrol mechanisms. There are commercially available examples of biocontrol agents.

Pathogens often develop resistance to the antibiotics and other mechanisms of biocontrol, so that they cannot be fully controlled in the long-term. A holistic approach with multiple controlling methods is probably better than excessive dependency on a single solution when confronting pathogens. Over the long term, pathogen-antagonistic bacteria will also evolve their mode of action to counteract the pathogens. PGPR also produce antibiotics such as lipopeptides, polyketides and antifungal metabolites that suppress pathogens.

Activity and Variety of Soil Microorganisms

The soil is the most outer layer of lithosphere. It covers almost the entire surface of the land; in it, there is a circulation of nutrients, biotic, and abiotic processes. Soil is a key element of the environment. It performs many functions in the environment. The five main ones are listed below:

- Environmental formation: Shaping the local climate, relief of the earth's surface, natural reservoir of water resources retention.
- Ecological: An element of circulation of biogenic elements and organic matter retention.
- Edaphic: Creation of living conditions for organisms (microorganisms, plants, animals.)

- Protection: Counteracting changes in the environment by means of sorption properties, e.g., absorbing toxic compounds.
- Economic: Place of work for man, a farmer.

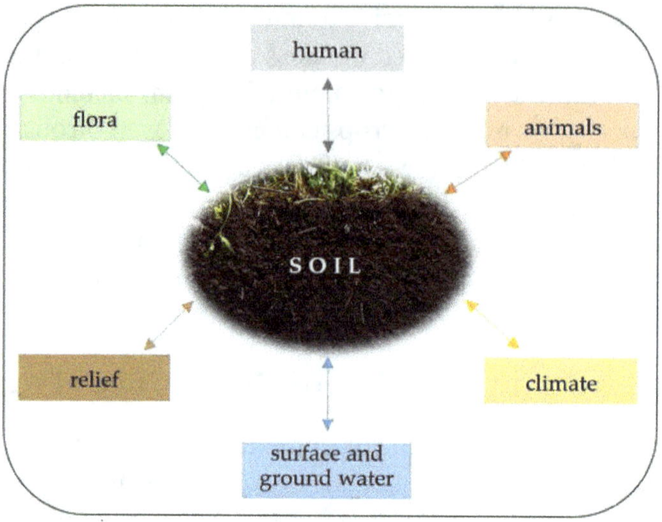

Soil functions in environment.

The soil has the capacity to self-produce resources necessary for the development of living organisms. This makes it a living environment for over 30% of the species existing on Earth. Soil organisms are a very important component of soils and are referred to as "biological engine of the earth". Soil quality can be defined as the balance between high activity and high microbiological biodiversity. Soil quality plays an important role in protecting the environment, preserving biodiversity and good agricultural practices. Agriculture has an impact on soil, and thus on the qualitative and quantitative composition of soil microorganisms.

Impact of Agricultural Activity on the Soil Quality

Human activities in the natural environment, above all the intensification of agriculture and the use of plant protection products, change the activity and diversity of the soil environment. This can lead to changes in the functioning and sustainability of the ecosystem. Changes in soil quality lead to changes in plant variety and productivity. Human activity is aimed at giving the soil the best possible soil production properties in order to increase the yield of plants.

Proper agricultural management should take into account the microbiological and physicochemical properties of the soil. The activity of microorganisms and their diversity is a sensitive indicator of soil quality, which is the subject of research for many researchers. The cultivation systems applied by man modify the quantitative and qualitative composition of organisms inhabiting the soil environment. It was found that soil microorganisms are sensitive to anthropogenic factors, in particular agricultural activity.

Numerous studies indicate that agricultural treatments strongly change chemical and physical parameters of soil. Changes in soil physiochemical parameters can lead to changes in soil microbial diversity. Some researchers suggest that the agricultural techniques generate a homogeneous soil environment and reduce bacterial diversity. Other studies indicate that the soil has a high capacity to maintain biodiversity. Numerous studies are being carried out to compare the impact of soil-tillage systems on soil microbiology. Some studies confirm that agricultural practices have an impact on the soil composition of microorganisms. Other researchers suggest that changes in the structure of the bacterial population are mainly due to chemical properties and soil structure. The research also reports that the soil has buffering capacity and that composition and functionality of microorganisms depends on the type of soil.

It is well known that the biodiversity of microorganisms varies from untapped soil to land cultivated in agricultural areas. In, it was demonstrated that in uncultivated soils, the obtained number of operational taxonomic units was about 1.5 million, 30% higher than agricultural soils. Similar results were obtained in which it was noted that 140–150 species per 1 g of soil are found in the soil cultivated for agriculture, while in natural soils, this amount is about 1000 species per 1 g of soil. It is estimated that about 40% of the world's agricultural land has already been degraded.

The Ecological Farming Systems

Sustainable agriculture is based on supporting natural biological processes without interfering with the environment. In simplified systems, the cultivation procedures are minimised. They consist mainly in loosening of the external soil layer, leaving crop residues and the use of stubble intercrops. Leaving harvest residues has an impact on soil nutrient management, which can be beneficial for the biological properties of soil. Among the organic systems, there is also the use of no-tillage, which is limited to the creation of a seed groove for sowing. Studies have shown that simplifications lead to an increase in the number of microorganisms of the genera Cladosporium and Mucor, which may inhibit the development of pathogens in soil. In addition, two to three times higher biomass of earthworms in the soil under direct sowing was recorded. The beneficial effect of organic fertilisation on soil microbiology is well known and documented.

The Conventional and Intensive Farming Systems

Traditional cultivation systems are based on specialised agricultural machinery that directly interferes with the soil structure. The deep ploughing up to 35 cm deep into the soil can cause negative consequences in the physical, chemical, and biological properties of the soil. Numerous studies have shown that conventional soil tillage and monoculture can have a negative impact on soil quality. Intensive ploughing leads to a more compact soil, which may result in lower yields. Intensive plant production and multiarea farms very often fail to take environmental protection into account and can have a detrimental effect on soil quality. In intensive agriculture, many mineral fertilisers, antibiotics, and hormones are used, which leads to degradation of the soil environment. At the same time, it carries a risk of biodiversity loss. The intensive mixing of the soil with the spacing plough causes losses of organic matter by accelerating the humus mineralization process. Such systems aggravate the network of interactions between soil organisms and aggravate disease-related problems. Many years of conventional long-term soil cultivation may also contribute to the decline of microbial population in soil.

Methods for Analysing the Impact of Human Agricultural Activities on Soil Quality

Research on soil quality is very important in terms of the quality of the yields obtained and the environmental impact assessment of agriculture. Soil as a habitat for many organisms and a place where many biochemical processes occur is sensitive to natural and abiotic factors, including agricultural activities. Soil quality assessment is based on an analysis of the physical-chemical and microbiological parameters of the soil.

Table: Selected soil quality parameters and methods of their evaluation.

Parameter	Method
Soil microbial biomass carbon (MBC) and nitrogen (MBN) contents.	The fumigation-extraction (F-E) and fumigation-incubation (F-I) method.
Potentially mineralizable nitrogen (PMN).	The short-term (waterlogged) anaerobic incubation method.
Dehydrogenases activity.	Method with 2,3,5-triphenyltetrazolium chloride (TTC) as a substrate.
Acid and alkaline phosphatases activity.	Method with p-nitrophenyl phosphate (PNP) as a substrate.
Urease activity.	Measuring the concentration of an ammonium.
3,6-diacetylfluorescein (FDA) hydrolysis activity.	Measuring the concentration of fluorescein.
Soil organic carbon (SOC).	Oxidation with a mixture of potassium dichromate and sulphuric acid.
Strength of soil aggregates.	Strength testing device.
Community level physiological profiling.	Biolog ECOplate.

Research on soil quality analysis:

- Determination of enzymatic activity, e.g., dehydrogenases, acidic and alkaline phosphatase, protease.
- Identification of soil microorganisms by molecular and immunological means through sequencing, hybridisation analysis, Elisa test, PCR-based reactions.
- Analyses of soil physical-chemical parameters e.g., pH, EC, granulometric composition, water volume.

- Evaluation of functional diversity of soil microorganisms by the CLPP method—Community Level Physiological Profiling.

However, despite many studies and the variety of methods used, there is still no single, universal parameter that would allow the quality of the soil to be determined quickly. In addition, it is necessary to carry out long-term studies, because only in this way can changes in the environment be noticed.

Impact of different Farming Systems on Selected Soil Quality Parameters

The Enzymatic Activity

Enzyme activity is considered to be one of the main parameters expressing the overall microbiological activity of the soil. The activity of dehydrogenase, peroxidase, phosphatase, protease and urease, β-glucosidase, catalase, cellulase, and invertase is commonly indicated. These enzymes are mediators and catalysts in many biochemical processes in soil, including formation and decay of humus, molecular nitrogen fixation, nitrification and denitrification, phosphorous compounds transformations, or detoxification of xenobiotics. The presence of enzymes and their activity are evidence of soil metabolism, while being sensitive to environmental changes.

Simplified systems have a high dehydrogenase activity. Restriction of plough use has a positive effect on the activity of these enzymes. In direct sowing soils, dehydrogenases also have a higher dehydrogenase activity than conventional crops.

Phosphatases catalyse the hydrolysis of organic phosphorus compounds once again responsible for its management in plants. They can be used as an indicator of the potential rate of phosphorus compounds mineralization. They are also very sensitive to heavy metal contamination. For acid phosphatase, higher values are recorded in soils from conventional crops and monoculture, e.g., the United States of America. Areas of the European Union's main agricultural holdings are covered by the so-called maize and alkaline phosphatase in simplified crops.

The FDA hydrolysis is related to the transformation of organic matter into soil. Many studies have shown that its value is higher in soils than in simplified, organic crops.

Urease activity is also sensitive to the way soil tillage is cultivated. It participates in the conversion of nitrogen compounds in soil. In, it was showed that the soils from simplified crops show up to two times higher urease activity than those from conventional crops.

The Soil Organic Carbon Content (SOC)

Soil organic carbon content is a sensitive soil quality indicator. The soil microbial biomass carbon (MBC) and nitrogen (MBN) represent a small fraction of the total soil organic C (1–5%) and N (2–6%).

Intensive monoculture of rice causes a decrease in SOC. Less intensive soil tillage improves the stability of soil aggregates and increases the SOC concentration. Researchers also observe higher SOC values in soil, where ploughing is not used and plant residues are left on the surface. In the case of maize, researchers have found that conventional cultivation reduces the amount of SOC in soil compared to zero tillage. Similar results were obtained in the case of winter wheat cultivation, where the SOC content in the upper soil layers was up to 23% higher than in the case of conventional seedlings. Conventional cultivation increases SOC sequestration. Research has shown that in agricultural soil can contain 20% less SOC and MBC than in soil under indigenous grassland.

The Microbial Biomass Carbon and Nitrogen Content

Determination of the soil microbiological biomass content is one of the basic parameters for monitoring soil changes. Soil management has a large impact on the size of the biomass pool of microorganisms.

There is a negative effect of conventional crops on the carbon and nitrogen content of microbial biomass. Whereas, the soil under simplified and organic cultivation has a high MBC and MBN content. The MBC and MBN in direct sowing soil are also higher than in conventional sowing. In several years of research, the number of MBC in conventional soil cultivation decreased, with a steady level of biomass in reduced tillage and direct sowing.

The Functional Analysis

Soil microorganisms are complex, multi-species communities. The processes taking place with their participation should be analysed as a sum of functions of the entire population and not as individual species. Using the Biolog Ecoplate, the observed a greater metabolic diversity in soil than in reduced tillage compared to conventional tillage. Similar results were obtained by. In soils from urban ecosystems, the highest microbiological activity was observed in roadside-tree soils, while the lowest activity was observed in forest sites. At the same time, the researchers did not notice any significant differences among long-term maize production, park sites, and street green sites.

The higher average utilisation intensity (AWCD) values were recorded in soil under conventional tillage with residue retention, than zero tillage and residue removal, after 15 years of cultivation. At the same time, it has been shown that the higher functional diversity of microorganisms is in soil under wheat cultivation compared to soil under maize cultivation. Research showed that ability of the microbial communities to utilise substrates on ECOplate was affected by fertilisation, but not by crop rotation treatments. At the same time, it has been demonstrated that the long-term use of manure can contribute to increase the beneficial functional diversity of microorganism in agricultural soils. It was demonstrated that the functional diversity of microbial community depended on the presence of plant species. The CLPP analysis also showed that the soil under cultivation of transgenic rice did not differ significantly in functional diversity of microbial communities compared to control.

The Biodiversity in Soils with Different Uses

The agricultural activity of mankind affects many of the physical and chemical properties of soil and its microbiological activity. Depending on the manner of soil utilisation, there are differences in the composition of the population of microorganisms inhabiting the soil. They concluded that the species diversity of soil microorganisms is primarily influenced by the abiotic soil properties independent of location or land-use type. They demonstrated that location and land-use type together explained more than 33% of the variation in soil biota diversity. However, they have shown that soils with low human participation rate (cork-oak, forest, and pasture) were characterised by a more stable bacterial environment than soils with a high human participation rate (vineyards and managed meadow). Research also shows that soils used for agricultural use are degraded, which is noted in lower microbiological parameters when compared with uncultivated soils. Changes in soil communities and loss of soil biodiversity threaten the multifunctionality and sustainability of the ecosystem.

Large variations in Proteobacteria composition are observed in soils with different uses. There were also changes in the composition of bacterial communities in the case of a change in the way forest soil are used to cultivate crops and grazed pastures. Studies conducted in Poland showed that in soils used for agricultural use, there was a significantly lower amount of ammonifying bacteria in comparison with soils not used for agricultural purposes. Similar results were obtained in other studies on agricultural soils.

It is well known that more intensively managed soils often contain less fungi biomass. Lack of ploughing while leaving plant residues on the soil surface may increase the amount of fungi pathogenic to plants (Fusarium sp.). At the same time, a long-term economy without ploughing leads to a deep diversification of microbial communities and also develops high fungal biomass in surface soil. Other researchers have shown that cultivation practices had little impact on the diversity of microbial communities.

Nitrogen fertilisation of soil influences the number of individual groups of microorganisms in soil. High nitrogen doses may lead to the accumulation of toxic substances, e.g., ammonia in soil, which does not have a beneficial effect on the Actinobacteria.

Collections of microorganisms decomposing cellulose and hemicellulose change the composition of the population in response to changes in the composition of organic fertilisers. The use of manure as a fertiliser in rice fields increased the activity of microorganisms, with a simultaneous decrease in this activity in the case of chemical fertilisers.

Higher mite density, Collembola and Myriapoda, were observed in forest ecosystems, but on grasslands more earthworms were observed.

The examples given above show that soil is a complex ecosystem, and the cultivation method has an impact on many soil parameters (biological, chemical, physical), and thus also the organisms that inhabit it. Changes in populations of soil organisms may have an impact on plant growth and thus on yields.

References

- What-is-Sustainable-Agriculture: westernsare.org, Retrieved 19 June, 2020
- Sustainable-agriculture, environment: nationalgeographic.com, Retrieved 10 April, 2020
- Kuypers, MMM; Marchant, HK; Kartal, B (2011). "The Microbial Nitrogen-Cycling Network". Nature Reviews Microbiology. 1 (1): 1–14. doi:10.1038/nrmicro.2018.9. PMID 29398704
- Organic-matter-and-microorganisms-in-soil-soil-microbiology, soil-microbiology-55525: biologydiscussion.com, Retrieved 09 January, 2019
- Lavelle, P.; Dugdale, R.; Scholes, R.; Berhe, A. A.; Carpenter, E.; Codispoti, L.; et al. (2005). "12. Nutrient cycling" (PDF). Millennium Ecosystem Assessment: Objectives, Focus, and Approach. Island Press. ISBN 978-1-55963-228-7. Archived from the original (PDF) on 2007-09-28
- Microorg: compost.css.cornell.edu, Retrieved 25 August, 2020
- Caranto, Jonathan D.; Lancaster, Kyle M. (2017-08-01). "Nitric oxide is an obligate bacterial nitrification intermediate produced by hydroxylamine oxidoreductase". Proceedings of the National Academy of Sciences. 114 (31): 8217–8222. doi:10.1073/pnas.1704504114. PMC 5547625. PMID 28716929

Soil Microorganisms 3

- **Bacteria**
- **Harmful Bacteria in Soil**
- **Fungi**
- **Algae**
- **Protozoa**

Soil microorganisms can be categorized into bacteria, actinomycetes, fungi, algae and protozoa. Each of these categories has different characteristics based upon the functions and properties of soil. This chapter closely examines all the aspects concerned with soil microorganisms to provide an extensive understanding of the subject.

Bacteria

Microbes in the soil are directly tied to nutrient recycling especially carbon, nitrogen, phosphorus and sulfur. Bacteria are a major class of microorganisms that keep soils healthy and productive.

Bacteria Characteristics

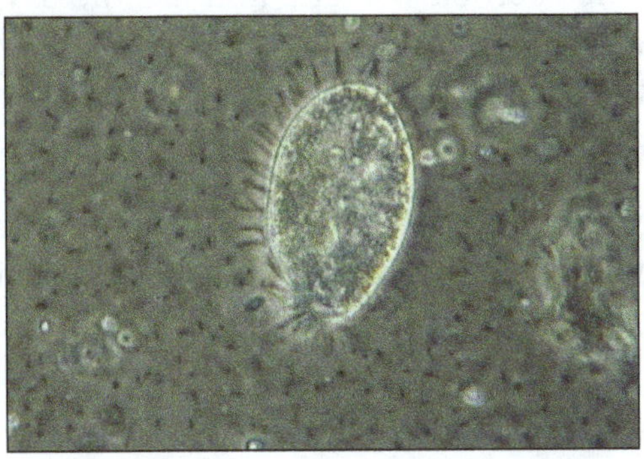

Close up view of a ciliate (protozoa) with various bacteria in the background.

"Bacteria are tiny one-celled organisms generally 4/100,000 of an inch wide (1 μm). A teaspoon of productive soil generally contains between 100 million and 1 billion bacteria. That is as much mass as two cows per acre. A ton of microscopic bacteria may be active in each acre." While bacteria may be small, they make up both the largest number and biomass (weight) of any soil microorganism.

Bacteria are similar in size to clay soil particles (<.2 μm) and silt soil particles (2-50 μm). They grow and live in thin water films around soil particles and near roots in an area called the rhizosphere. Bacteria's small size enables them to grow and adapt more rapidly to changing environmental conditions than larger, more complex microorganisms like fungi.

Most soils are simply a graveyard for dead bacteria cells. Bacteria are so simple in structure that they have often been called a bag of enzymes and soluble bags of fertilizer. Since bacteria live under starvation conditions or soil water stress, they reproduce quickly when optimal water, food, and environmental conditions occur. Bacteria population may easily double in 15-30 minutes. Flourishing microbial populations increase soil productivity and crop yields over time.

Bacteria Classification

Most bacteria are classified into one of the following four categories.

Bacteria Shape

When scientists started first classifying bacteria, they started by looking at their basic shape. Bacteria generally have three major shapes: rod, sphere or spiral. Actinomycetes are still classified as bacteria but are similar to fungi except they are smaller in size. Classifying bacteria by shape is complex because many bacteria have different shapes and different arrangements.

Aerobic and Anaerobic Bacteria

Most microbes are generally inactive and may only have short burst of soil activity. Soil oxygen levels often determine soil bacteria activity. Most soil bacteria prefer well-oxygenated soils and are called aerobic bacteria and use the oxygen to decompose most carbon compounds. Examples of aerobic bacteria include the Aerobacter genus which is widely distributed in the soil and actinomycetes bacteria genus Streptomyces which give soil its good "earthy" smell.

Anaerobic bacteria prefer and some require an environment without oxygen. Anaerobic bacteria are generally found in compacted soil, deep inside soil particles (microsites), and hydric soils where oxygen is limiting. Many pathogenic bacteria prefer anaerobic soil conditions and are known to outcompete or kill off aerobic bacteria in the soil. Many anaerobic bacteria are found in the intestines of animals and are associated with manure and bad smells.

Gram Negative and Gram Positive Bacteria

When a staining agent is used in the lab, bacteria can be classified as "gram negative" or "gram positive." The staining agent attaches to the bacteria's cell walls. Gram negative bacteria are generally the smallest bacteria and are sensitive to drought and water stress. Gram positive bacteria are much larger in size, have thicker cell walls, negative charges on the outside cell wall surface and tend to resist water stress. Bacteroides are anaerobic gram negative bacteria that live in the gut of humans and animals. Listeria is a gram positive aerobic rod shaped bacteria found in contaminated food.

Other Bacteria Classifications

Another way to classify bacteria is by their growth and reproduction. Autotrophic bacteria (also called autotrophs) process carbon dioxide to get their carbon. Some autotrophic bacteria directly use sunlight and carbon dioxide to produce sugars, while others depend on other chemical reactions to obtain energy. Algae and cyanobacteria are some examples of autotrophic bacteria. Heterotrophic bacteria obtain their carbohydrates and sugars from their environment or the living organism or cell they inhabit. Examples include Arthrobacter bacteria involved in nitrogen nitrification.

With new advances in DNA sequencing, most scientists are classifying bacteria based on the type of environment in which they inhabit. Bacteria can live in extreme environments like hot springs for sulfur bacteria or in extreme cold as in ice water in the Arctic. Bacteria may also be classified by living in a highly acidic versus alkaline environment, aerobic versus anaerobic, or autotrophic versus heterotrophic environment.

Bacteria Functional Groups

Bacteria perform many important ecosystem services in the soil including improved soil structure and soil aggregation, recycling of soil nutrients, and water recycling. Soil bacteria form microaggregates in the soil by binding soil particles together with their secretions. These microaggregates are like the building blocks for improving soil structure. Improved soil structure increases water infiltration and increases water holding capacity of the soil.

Bacteria perform important functions in the soil, decomposing organic residues from enzymes released into the soil. Ingham describes the four major soil bacteria functional groups as decomposers, mutualists, pathogens and lithotrophs. Each functional bacteria group plays a role in recycling soil nutrients.

The decomposers consume the easy-to-digest carbon compounds and simple sugars and tie up soluble nutrients like nitrogen in their cell membranes. Bacteria dominate in tilled soils but they are only 20-30 percent efficient at recycling carbon (C). Bacteria are higher in nitrogen (N) content (10-30 percent nitrogen, 3 to 10 C:N ratio) than most microbes.

Of the mutualistic bacteria, there are four bacteria types that convert atmospheric nitrogen (N_2) into nitrogen for plants. There are three types of soil bacteria that fix nitrogen without a plant host and live freely in the soil and these include Azotobacter, Azospirillum and Clostridium.

Nitrogen fixing Rhizobium bacteria form nodules on a soybean root.

The Rhizobium bacteria (gram negative rod-shaped bacteria) species associate with a plant host: Legume (alfalfa, soybeans) or clover (red, sweet, white, crimson) to form nitrogen nodules to fix nitrogen for plant growth. The plant supplies the carbon to the Rhizobium in the form of simple sugars. Rhizobium bacteria take nitrogen from the atmosphere and convert it to a form the plant can use. For plant use, the atmospheric nitrogen (N_2) or reactive nitrogen combines with oxygen to form nitrate (NO_3^-) or nitrite (NO_2^-) or combines with hydrogen to produce ammonia (NH_3^+) or ammonium (NH_4^+) which are used by plant cells to make amino acids and proteins.

Many soil bacteria process nitrogen in organic substrates, but only nitrogen fixing bacteria can process the nitrogen in the atmosphere into a form (fixed nitrogen) that plants can use. Nitrogen fixation occurs because these specific bacteria produce the nitrogenase enzyme. Nitrogen fixing bacteria are generally widely available in most soil types (both free living soil species and bacteria species dependent on a plant host). Free living species generally only comprise a very small percentage of the total microbial population and are often bacteria strains with low nitrogen fixing ability.

Nitrification is a process where nitrifying bacteria convert ammonia (NH_4^+) to nitrite (NO_2^-) and then to nitrate (NO_3^-). Bacteria and fungi are typically consumed by protozoa and nematodes and the microbial wastes they excrete is ammonia (NH_4^+) which is plant available nitrogen. Nitrite bacteria (Nitrosomonas spp.) convert the ammonia into nitrites (NO_2^-) and nitrate bacteria (Nitrobacter spp.) may then convert the nitrites (NO_2^-) to nitrates (NO_3^-). Nitrifying bacteria prefer alkaline soil conditions or a pH above 7 is. Both nitrate and ammonia are plant available forms of nitrogen; however, most plants prefer ammonia because the nitrate has to be converted to ammonia in the plant cell in order to form amino acids.

Denitrifying bacteria allow nitrate (NO_3^-) to be converted to nitrous oxide (N_2O) or dinitrogen (N_2) (atmospheric nitrogen). For denitrification to occur, a lack of oxygen or anaerobic conditions must occur to allow the bacteria to cleave off the oxygen. These conditions are common in ponded or saturated fields, compacted fields, or deep inside the microaggregates of soil where oxygen is limited. Denitrifying bacteria decrease the nitrogen fertility of soils by allowing the nitrogen to escape back into the atmosphere. On a saturated clay soil, as much as 40 to 60 percent of the soil nitrogen may be lost by denitrification to the atmosphere.

Pathogenic bacteria cause diseases in plants and a good example are bacteria blights. Healthy and diverse soil bacteria populations produce antibiotics that protect the plants from disease causing organisms and plant pathogens. Diverse bacteria populations compete for the same soil nutrients and water and tend to act as a check and balance system by reducing the disease-causing organism populations. With high microbial diversity, soils have more nonpathogenic bacteria competing with the pathogenic bacteria for nutrients and habitat. Streptomycetes (actinomycetes) produce more than 50 different antibiotics to protect plants from pathogenic bacteria.

Lithotrophs (chemoautotrophs) get their energy from compounds other than carbon (like nitrogen or sulfur) and include species important in nitrogen and sulfur recycling. Under well-aerated conditions, sulfur-oxidizing bacteria make the sulfur more plant available while under saturated (anaerobic, low oxygen) soil conditions, sulfur reducing bacteria make sulfur less plant available.

Actinomycetes have large filaments or hyphae and act similar to fungus in processing soil organic residues which are hard to decompose (chitin, lignin, etc.). When farmers plow or till the soil, actinomycetes release "geosmin" as they die which gives freshly turned soil its characteristic smell. Actinomycetes decompose many substances but are more active at high soil pH levels. Actinomycetes are important in forming stable humus, which enhances soil structure, improves nutrient storage, and increases water retention.

Soil Benefits from Bacteria

Bacteria grow in many different microenvironments and specific niches in the soil. Bacteria populations expand rapidly and the bacteria are more competitive when easily digestible simple sugars are readily available around in the rhizosphere. Root exudates, dead plant debris, simple sugars, and complex polysaccharides are abundant is this region. About 10 to 30 percent of the soil microorganisms in the rhizosphere are actinomycetes, depending on environmental conditions.

Many bacteria produce a layer of polysaccharides or glycoproteins that coats the surface of soil particles. These substances play an important role in cementing sand, silt and clay soil particles into stable microaggregates that improve soil structure.

Bacteria live around the edges of soil mineral particles, especially clay and associated organic residues. Bacteria are important in producing polysaccharides that cement sand, silt and clay particles together to form microaggregates and improve soil structure. Bacteria do not move very far in the soil, so most movement is associated with water, growing roots or hitching a ride with other soil fauna like earthworms, ants, spiders, etc.

In general, most soil bacteria do better in neutral pH soils that are well oxygenated. Bacteria provide large quantities of nitrogen to plants and nitrogen is often lacking in the soil. Many bacteria secrete enzymes in the soil to makes phosphorus more soluble and plant available. In general, bacteria tend to dominate fungi in tilled or disrupted soils because the fungi prefer more acidic environments without soil disturbance. Bacteria also dominate in flooded fields because most fungi do not survive without oxygen. Bacteria can survive in dry or flooded conditions due to their small size, high numbers, and their ability to live in small microsites within the soil where environmental conditions may be favorable. Once the environmental conditions around these microsites become more favorable, the survivors quickly expand their populations. Protozoa tend to be the biggest predators of bacteria in tilled soils.

In order for bacteria to survive in the soil, they must adapt to many microenvironments. In the soil, oxygen concentrations vary widely from one microsite to another. Large pore spaces filled with air provide high levels of oxygen, which favors aerobic conditions, while a few millimeters away, smaller micropores may be anaerobic or lack oxygen. This diversity in soil microenvironments allows bacteria to thrive under various soil moisture and oxygen levels, because even after a flood (saturated soil, lack of oxygen) or soil tillage (infusion of oxygen) small microenvironments exist where different types of bacteria and microorganisms may live to repopulate the soil when environmental conditions improve.

Natural succession happens in a number of plant environments including in the soil. Bacteria improve the soil so that new plants can become established. Without bacteria, new plant populations and communities struggle to survive or even exist. Bacteria change the soil environment so that certain plant species can exist and proliferate. Where new soil is forming, certain photosynthetic bacteria start to colonize the soil, recycling nitrogen, carbon, phosphorus, and other soil nutrients to produce the first organic matter. A soil that is dominated by bacteria usually is tilled or disrupted and has higher soil pH and nitrogen available as nitrate, which is the perfect environment for low successional plants called weeds.

As the soil is disturbed less and plant diversity increases, the soil food web becomes more balanced and diverse, making soil nutrients more available in an environment better suited to higher plants. Diverse microbial populations with fungus, protozoa and nematodes keep nutrients recycling and keep disease-causing organisms in check.

Actinomycetes

Actinomycetes are a remarkable type of bacteria. So far we know they function in at least three important ways.

- The Streptomyces family, provides us with many of the important antibiotics we use in medicine today.
- The Frankia family, works in a symbiotic relationship with many non legume plants as nitrogen fixing bacteria.
- Other actinomycetes are decomposers digesting tough plant tissues like the cellulose and lignin in bark, paper and stems and the chitin or hard exoskeletons of insects.

Actinomycetes are a fungi-like bacteria forming long filaments that stretch through the soil. They have sometimes been classed as fungi because they both look like, and decompose similar material as fungi. However, they have no defined nucleus. As well antibacterial agents work against them but antifungal agents do not. So now we can call them actinobacteria.

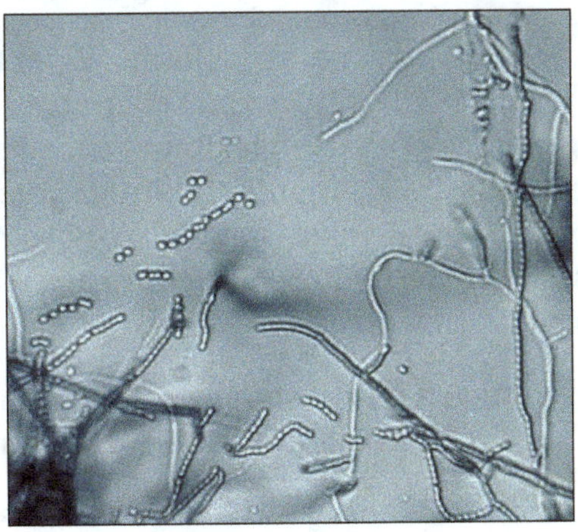

Streptomyces spp.

Scientists have long known that actinomycetes somehow kept soil bacteria populations in balance. This fact led scientists Selman Waksman and Albert Shatz to think that some of these actinobacteria might be useful as antibiotics against human disease. Shatz was particularly driven by his experience as a lab assistant at a Miami Hospital during the 2nd World War. Here he witnessed the death of many young soldiers at the hand of penicillin immune infections, especially Tuberculosis.

Shatz, when released from the Miami Hospital returned to Dr Selman's lab at Rutger's University. On October 19th, 1943, after three months of work, he managed to isolate two strands of actinobacteria that could effectively stop the growth of penicillin resistant bacteria. Streptomycin, the first antibiotic to cure tuberculosis, was the result of this work. Streptomycin is still used, but in combination with other drugs, in the battle against TB.

The Streptomyces produce over two-thirds of the clinically useful antibiotics that are natural in origin including Erythromycin, Neomycin, Tetracycline and Cefoxitin. Kind of puts mud baths and masks along with the age old childhood habit of eating dirt in a different light, doesn't it.

Nitrogen Fixing Bacteria – Frankia

Root nodules with frankia nitrogen fixing antinobacteria.

Nitrogen is a critical nutrient for virtually all lifeforms. We get our nitrogen either directly or indirectly from plants. While nitrogen makes up about 79% of our atmosphere, plants can not use nitrogen in its gaseous state. It first must be fixed or combined into either ammonia, NH_3 or Nitrate, NO_3. The natural nitrogen cycle relies on nitrogen fixing bacteria like those found in the Frankia family of actinobacteria, to supply the fixed nitrogen. Fixed nitrogen is often the limiting factor for growth, both in your garden and in the general environment.

About 15% of the world's nitrogen fixed naturally is from symbiotic relationships between various species of the Frankia family of actinobacteria and their host plants.

The plants that form symbiotic relationships with Frankia are called actinorhizal plants. Scientists have found over 160 plants that host these actinomycetes including alders, Russian olive, bayberry, sweet fern, bitterbush and cliffrose. The Frankia is able to provide most or all of the host plant's nitrogen needs.

These nitrogen fixing bacteria and their host plants are often pioneer species on young nitrogen deficient and disturbed soils such as moraines, volcanic flows and sand dunes. They help create a reservoir of nitrogen rich soil that the next wave of plants can benefit from.

Scientists believe that much of the new nitrogen in temperate forests, dry chaparral, sand dunes, moraines, and mine waste tailings is thanks to the mutualism of Frankia and host plants. They are the main nitrogen fixing relationships in large parts of the world and will only become more important as we adjust to climate change.

Decomposer Actinomycetes

As a decomposer the actinomycetes specialize in breaking down tough cellulose and lignin found in wood and paper and the chitlin found in the exoskeletons of insects. The breakdown of these materials makes nutrients once again available to plants. The "earthy" smell of freshly turned soil is caused by actinomycetes at work.

While some species of actinobacteria are thermophiles, loving the heat of an active hot compost. During the long cool curing stage of compost they extend their hyphae like threads throughout the compost. They are those whitish grey cobweb like threads.

Nature and Habitat

Actinomycetes are the most abundant organisms that form thread-like filaments in the soil. They grow as hyphae like fungi responsible for the characteristically "earthy" smell of freshly turned healthy soil. The actinomycetes exist in various habits in nature and represent a ubiquitous group of microbes widely distributed in natural ecosystems around the world.

They are primarily soil inhabitants but have been found widely distributed in a diverse range of aquatic ecosystem, including sediments obtained from deep sea, even from greatest depth Mariana Trench. Their presence in extreme environments especially at cryophilic region, for example soil taken from Antarctica and even from desert soil has been reported. It has been demonstrated in a comparative survey that actinomycete population is largest in soils of surface layer and gradually decreases with as depth increased; individual actinomycete strains are present in all soil layers.

Structure

Actinomycetes are characterized by the formation of normally branching threads or rods. The hyphae are generally non-septate; under certain special conditions, septa may be observed in some forms. The sporulating mycelium may be branching or non-branching, straight or spiral shaped. The spores are spherical, cylindrical or oval. Actinomycetes produce initial micro colonies composed of branching system filaments that after 24-48 hours fragment into diptheroids, short chain and coccobacillary forms. The cell wall of actinomycetes is a rigid structure that maintains the shape of the cell wall of actinomycetes that maintains the shape of the cell ad prevents bursting of the cell due to high osmotic pressure.

The wall consists of a large variety of complex compounds including peptidoglycan, teichoic and teichuronic acid and polysaccharides. The peptidoglycan consists of glycan (polysaccharides) chains of alternating N-acetyl-d-glucosamine (NAG) and N-acetyl-d-muramic acid (NAM) and diaminopimelic acid (DAP), which is uique in prokaryotic cell walls. Teichoic and teichuronic acid are chemically bonded to peptidoglycan.

The chemical composition of their cell wall is similar to that of gram positive bacteria but because of their well developed morphological (hyphae) and cultural characteristics, actinomycetes have been considered as a group, well separated from other common bacteria.

Taxonomic Outline of Phylum Actinobacteria

Actinomycetes are traditionally classified as part of the bacteria. In the Bergey's Manual of Determinative Bacteriology, Actinomycetes are included in several sections of volume four. All Actinomycetes are included under the order Actinomycetales. The order Actinomycetales is divided in to four families- Streptomycetaceae, Actinomycetaceae, Actinoplanaceae, and Mycobacteriaceae. The "Bergey's Manual of Systematic Bacteriology-2nd edition" for actinomycetes classification has five volumes, that contain internationally recognized names and descriptions of bacteria species. Classification of actinomycetes has been rearrange d as follows:

"Bergey's Manual of Systematic Bacteriology"

Volume 1 (2001)	Volume 2 (2005)	Volume 3 (2009)	Volume 4 (2011)	Volume 5 (2012)

Volume.1: The Archaea and the deeply branching and phototrophic Bacteria.
Volume.2: The Proteobacteria.
Volume.3: The Firmicutes.
Volume.4: The Bacteroidetes, Spirochaetes, Tenericutes (Mollicutes), Acidobacteria, Fibrobactres, Fusobacteria,
Volume.5: The Actinobacteria

In volume 5, the phylum actinobacteria is divided into 6 classes namely Actinobacteria, Acidimicrobiia, Coriobacteriia, Nitriliruptoria, Rubrobacteria and Thermoleophilia. The class actinobacteria is further divided into 16 orders which are Actinomycetales, Actinopolysporales, Bifidobacteriales, Catenulisporales, Corynebacteriales, Frankiales, Glycomycetales, Jiangellales, Kineosporiales, Micrococcales, Micromonosporales, Propionibacteriales, Pseudonocardiales, Streptomycetales, Streptosporangiales,Incertae sedis.

The family Actinomycetaceae (under the order Actinomycetales) and Streptomycetaceae (under the order Streptomycetales) comprise those actinomycetes, with which this treatise is primarily concerned.

Actinomycetes have also been classified into several groups based on biochemical parameters. Based on major cell wall constituent, four groups have been identified in Actinomycetes.

Cell Wall Type	Sugar Pattern	Genera
	No characteristic sugar pattern.	Streptomyces, Streptoverticillicum etc.

	Araginose, Xylose (monosaccharide).	Actinoplanes, Micromonospora etc.
	No sugar.	Dermatophilus, Planomonospora stc.
	Galactose, Arabinose.	Mycobacterium, Nocardia etc.

Significance of Actinomycetes

Actinomycetes are biotechnologically valuable bacteria which are well exploited for secondary metabolites. Screening, isolation and characterization of promising strains of actinomycetes producing potential secondary metabolites has been a major area of research by many groups worldwide for many years. Among various genera of Actinomycetes; Streptomyces, Saccharopolyspora, Amycolatopsis, Micromonospora and Actinoplanes are the major producers of commercially important biomolecules. Actinomycetes especially Streptomyces species are widely recognized as industrially important microorganisms as they are a rich source of several useful bioactive natural products with potential applications and are prolific producers of secondary metabolites, many of which have commercial importance as antibiotics, anti-parasitics and antifungal agents, herbicides, pesticides, anticancer or immunosuppressive agents as well as industrially important enzymes. Of all known drugs, 70% have been isolated from Actinomycetes bacteria and out of which 75% and 60% are used in medicine and agriculture respectively. Secondary metabolites, to a large extent, are species-specific and in contrast to primary metabolites, often accumulate in substantial quantity, which is a key factor in their commercial importance. Actinomycetes are of tremendous economic importance as the secondary metabolites produced by them includes antibiotics, other medicinals, toxins, pesticides and animal and plant growth factors. The best known of the secondary metabolites produced by Actinomycetes are the antibiotics.

Antibiotics are truly referred as the 'wonder drugs' for their virtual success against pathogenic microorganisms. This remarkable group of compounds forms a heterogeneous assemblage of biologically active molecules with different structures and modes of action. They attack virtually every type of microbial activity such as DNA, RNA, and protein synthesis, membrane function, electron transport, sporulation, germination and many others. As a result, they are effective treatments for bacterial infections. Prior to the discovery of antibiotics, people with simple wounds and infectious diseases could not be treated.

The first antibiotic discovered by Sir Alexander Fleming in 1928 facilitated the discovery of many other secondary metabolites with similar properties. The secondary metabolites produced by actinomycetes serve as the sources of life saving environments.These have a broad spectrum of biological activities; e.g. antibacterial (streptomycin, tetracycline, chloramphenicol), antifungal (nystatin), antiviral (tunicamycin), antiparasitic (avermectin), immunosuppressive (rapamycin), antitumor (actinomycin, mitomycinC, anthracyclines), enzyme inhibitory (clavulanic acid) and diabetogenic (bafilomycin, streptozotocin), Cancer (doxorubicins, daunorubicin, mitomycin and bleomycin), transplant rejection (cyclosporine and rapamycin) and high cholesterol (statins such as lovastatin and mevastatin.

Members of Actinomycetes group,in addition are producers of clinically useful antitumor drugs such as Anthracyclines (Aclarubicin, Daunomycin and Doxorubicin), Peptides (Bleomycin and Actinomycin D), Aurelic acids (Mithramycin), Enediynes (Neocarzinstatin), Antimetabolites (Pentostatin), Carzinophilin, Mitomycins and others. Actinomycetes products are not able not only for their potent therapeutic activities but also for the fact that they frequently possess the desirable pharmacokinetic properties required for clinical development.

The actinomycetes are important not only in the field of pharmaceutical industries and also in the agriculture. Actinomycetes have the potential to inhibit the growth of several plant pathogens e.g. Erwinia amylovora (bacteria that cause fireblight to apple), Agrobacterium tumefaciens.

Actinomycetes decompose complex mixtures of polymer in dead plant, animal and fungal material resulting in production of many extracellular enzymes which are conductive to crop production. Apart from this, Actinomycetes have major contribution in biological buffering of soils, biological control of soil environments by nitrogen fixationand degradation of high molecular weight compounds like hydrocarbons in the polluted soil.

Actinomycetes have the capacity to synthesize many different biologically active secondary metabolites such as cosmetics, vitamins, nutritional materials, herbicides, antibiotics, pesticides, anti-parasitic and enzymes like cellulose and xylanase used in waste treatment.

Actinomycetes also synthesize and excrete dark pigments, melanin or melanoid. These melanin compounds are irregular, dark brown polymers having radio protective and antioxidant properties that can effectively protect the living organisms from ultraviolet radiation. Melanins are frequently used in medicine, pharmacology, and cosmetics preparations.

Actinobacteria

Actinobacteria, which share the characteristics of both bacteria and fungi, are widely distributed in both terrestrial and aquatic ecosystems, mainly in soil, where they play an essential role in recycling refractory biomaterials by decomposing complex mixtures of polymers in dead plants and animals and fungal materials. They are considered as the biotechnologically valuable bacteria that are exploited for its secondary metabolite production. Approximately, 10,000 bioactive metabolites are produced by Actinobacteria, which is 45% of all bioactive microbial metabolites discovered. Especially Streptomyces species produce industrially important microorganisms as they are a rich source of several useful bioactive natural products with potential applications. Though it has various applications, some Actinobacteria have its own negative effect against plants, animals, and humans. On this context, this chapter summarizes the general characteristics of Actinobacteria, its habitat, systematic classification, various biotechnological applications, and negative impact on plants and animals.

Actinobacteria are a group of Gram-positive bacteria with high guanine and cytosine content in their DNA, which can be terrestrial or aquatic. Though they are unicellular like bacteria, they do not have distinct cell wall, but they produce a mycelium that is nonseptate and more slender. Actinobacteria include some of the most common soil, freshwater, and marine type, playing an important role in decomposition of organic materials, such as cellulose and chitin, thereby playing a vital part in organic matter turnover and carbon cycle, replenishing the supply of nutrients in the soil, and is an important part of humus formation. Actinobacterial colonies show powdery consistency and stick firmly to agar surface, producing hyphae and conidia/sporangia-like fungi in culture media.

Actinobacteria produce a variety of secondary metabolites with high pharmacological and commercial interest. With the discovery of actinomycin, a number of antibiotics have been discovered from Actinobacteria, especially from the genus Streptomyces. They are widely distributed in soil with high sensitivity to acid and low pH. Actinobacteria have a number of important functions, including degradation/decomposition of all sorts of organic substances such as cellulose, polysaccharides, protein fats, organic acids, and so on. They are also responsible for subsequent decomposition of humus (resistant material) in soil and for the earthy smell of freshly ploughed soils, producing a number of antibiotics like streptomycin, terramycin, aureomycin, and so on.

Habitat of Actinobacteria

Terrestrial Environment

Soil remains the most important habitat for Actinobacteria with streptomycetes existing as a major component of its population. According to numerous reports, Streptomyces was encountered to be the most abundant genus isolated in each of the study. Terrestrial Actinobacteria have various interesting antimicrobial potentials. Oskay et al Actinobacteria that had capability of producing novel antibiotics with high antibacterial activity. In anoxic mangrove rhizosphere, Actinobacterial species such as Streptomyces, Micromonospora, and Nocardioform were found to be abundant, which is 1000 to 10000 times smaller than arable lands because of tidal influence. Similarly, Nocardia isolated from mangrove soil produced new cytotoxic metabolites that strongly inhibited human cell lines, such as gastric adenocarcinoma. Dessert soil is also considered as an extreme terrestrial environment where only certain species, especially wherein Actinobacteria, often use Microcoleus as a source of food. There are several reports showing the distribution of Actinobacteria in various locations, such as sandy soil, black alkaline soil, sandy loam soil, alkaline dessert soil, and subtropical dessert soil, where Streptomyces sp. were dominant followed by the other organisms, such as Nocardia, Nocardiopsis, and Actinomycetes. In the study of Nithya et al, 134 morphologically distinguished culturable Actinobacteria were isolated from 10 different desert soil samples, and the isolates were found to have varying level of antibacterial activity against bacterial pathogens. Equally, Actinobacteria play a major part in rhizosphere microbial community in the turnover of recalcitrant plant organic matter, and thus the rhizosphere region is considered as one of the best habitats for isolation of these microorganisms. Priyadharsini et al in her study isolated 45 morphologically distinct colonies from 12 different paddy field soils and observed their ability to inhibit the growth of Cyperus rotundus. The isolates include Streptomyces sp., Streptoverticillium sp., Actinomadura sp., Kitasatosporia sp., Nocardiopsis sp., Pseudonocardia sp., and Kibdelosporangium sp.

General Characteristics of Actinobacteria

Actinobacteria comprises a group of branching unicellular microorganisms, most of which are aerobic-forming mycelium known as substrate and aerial. They reproduce by binary fission or by producing spores or conidia, and sporulation of Actinobacteria is through fragmentation and segmentation or conidia formation. The morphological appearance of Actinobacteria is compact, often leathery, giving a conical appearance with a dry surface on culture media and are frequently covered with aerial mycelium.

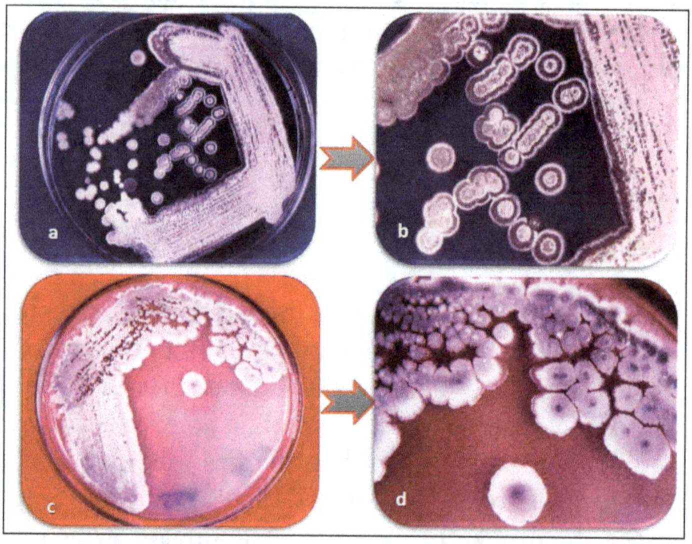

Appearance of Actinobacteria isolates on Starch casein agar plate. a, c Plate view of the Actinobacterial isolates. b, d Morphology of individual colonies.

Aerial Mycelium

The aerial mycelium is usually thicker than the substrate mycelium. The aerial mycelium shows sufficient differentiation that a miscellaneous assortment of isolates can be segregated into a number of groups having similar morphological characteristics under fixed condition. This is designated as one of the most important criteria for the classification of the genus Streptomyces into species, comprising structure (cottony, velvety, or powdery), formation of rings or concentric zones, and pigmentation.

Substrate Mycelium

The substrate mycelium of Actinobacteria varies in size, shape, and thickness. Its color ranges from white or virtually colorless to yellow, brown, red, pink, orange, green, or black.

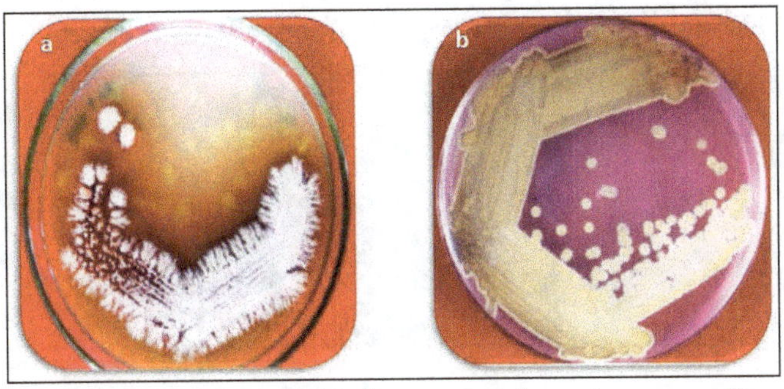

Abundant growth of Actinobacterial isolate on starch casein agar medium. a. Aerial mycelium. b. Reverse side of plate showing substrate mycelium.

Morphological Appearance

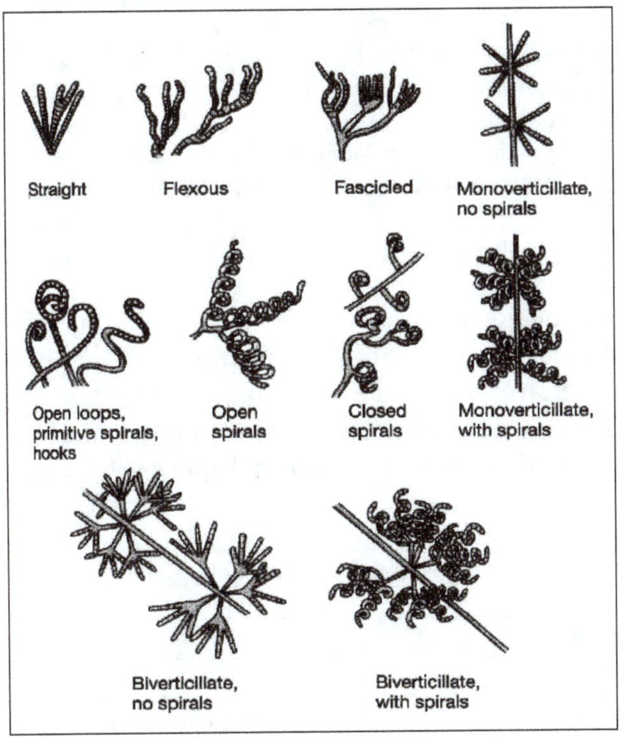

Type of spore-bearing structure in streptomycetes.

Morphology has been an important characteristic to identify Actinobacteria isolates, which was used in the first descriptions of Streptomyces species. This is made using various standard culture media, including International Streptomyces Project (ISP). For nonstreptomycetes or rare Actinobacteria, strains maintained on ATCC Medium No.172 (NZ-amine glucose starch agar) were used. Various morphological observations, including germination of spores, elongation and branching of vegetative mycelium, formation of aerial mycelium, color of aerial and substrate mycelium, and pigment production, have been used to identify Actinobacteria. Light microscopy was used to study the formation of aerial mycelium and substrate mycelium, and scanning electron microscopy was used to study the spores, the spore surface, and spore structure.

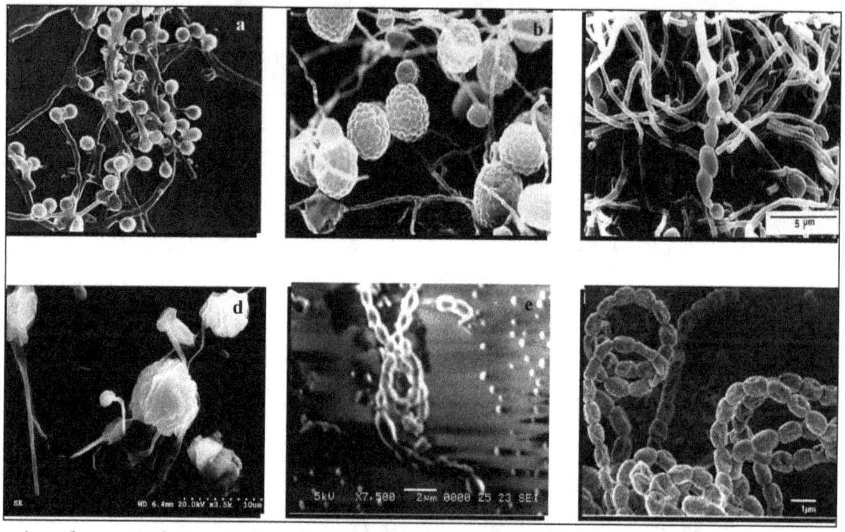

Scanning electron photographs of various Actinobacterial isolates. a. Micromonospora sp. b. Streptosporangium sp. c. Saccharopolyspora sp. d. Actinosynnema sp. e. S. noursei DPTD21. f. Streptomyces sp. JD9.

Systematics of Actinobacteria

All Actinobacteria are included under the order Actinomycetales. The order Actinomycetales is divided into four families—Streptomycetaceae, Actinomycetaceae, Actinoplanaceae, and Mycobacteriaceae. The "Bergey's Manual of Systematic Bacteriology—2nd edition" for Actinobacteria classification has five volumes, which contain internationally recognized names and descriptions of bacterial species. Classification of Actinobacteria has been rearranged as follows:

The Archaea and the Deeply Branching and Phototrophic Bacteria	Volume 1
The Proteobacteria	Volume 2
The Firmicutes	Volume 3
The Bacteroidetes, Spirochaetes, Tenericutes (Mollicutes), Acidobacteria, Fibrobacteres, Fusobacteria	Volume 4
The Actinobacteria	Volume 5

In Volume 5, the phylum Actinobacteria is divided into six classes, namely Actinobacteria, Acidimicrobiia, Coriobacteriia, Nitriliruptoria, Rubrobacteria, and Thermoleophilia. The class Actinobacteria is further divided into 16 orders that are Actinomycetales, Actinopolysporales, Bifidobacteriales, Catenulisporales, Corynebacteriales, Frankiales,

Glycomycetales, Jiangellales, Kineosporiales, Micrococcales, Micromonosporales, Propionibacteriales, Pseudonocardiales, Streptomycetales, Streptosporangiales, and Incertae sedis. In the order of abundance in soils, the common genera of Actinobacteria are Streptomyces (nearly 70%), Nocardia, and Micromonospora, although Actinoplanes, Micromonospora, and Streptosporangium are also generally encountered. At present, the molecular identification is based on 16S rDNA sequences, which is most significant for Actinobacteria identification.

Types of Actinobacteria

Thermophilic Actinobacteria

Number of studies has been carried out by the researchers to confirm the existence of extremophilic and extreme tolerant soil Actinobacteria (acid tolerant and alkali tolerant, psychrotolerant and thermotolerant, and halotolerant and haloalkalitolerant or xerophilic). Mesophilic Actinobacteria can grow at an optimal temperature from 20 °C to 42 °C, among which thermotolerant species exist, which can survive at 50 °C. Moderately thermophilic Actinobacteria have an optimum growth at 45 °C–55 °C, whereas strictly thermophilic Actinobacteria grow at 37 °C–65 °C with the optimum temperature at 55 °C–60 °C. Incubation temperatures of 28 °C, 37 °C, and 45 °C are considered optimal for isolation of soil mesophilic, thermotolerant, and moderately thermophilic Actinobacteria. Thermoactinomyces, which is presently excluded from the order Actinomycetales, are described as thermophilic forms depending on its phenotypic and molecular genetic characteristics, as well as among some species of Thermomonospora, Microbispora, Saccharopolyspora, Saccharomonospora, and Streptomyces.

Acidophilic Actinobacteria

Acidophilic Actinobacteria, which are common in terrestrial habitats such as acidic forest and mine drainage soil, grow in the pH range from about 3.5 to 6.5, with optimum rates at pH 4.5 to 5.5. It has been shown that acidophilic Actinobacteria consistently form two distinct aggregate taxa (namely, the neutrotolerant acidophilic and strictly acidophilic cluster groups) based on numerical phenetic data; members of the two groups share common morphological and chemotaxonomic properties. Also some members of the strictly acidophilic group form a distinct taxon, such as the genus Streptacidiphilus, which has been assigned to the revised family Streptomycetaceae, together with the genera Kitasatospora and Streptomyces.

Halophilic Actinobacteria

Halophilic Actinobacteria are categorized into different types based on their growth in media containing different concentrations of salt. Extreme halophiles grow best in media containing 2.5–5.2 M salt, whereas borderline extreme halophiles grow best in media containing 1.5–4.0 M salt, moderate halophiles grow best in media containing

0.5–2.5 M salt, and finally halotolerants that do not show an absolute requirement to salt for growth but grow well up to often very high salt concentrations and tolerate 100 g/l salt (equivalent to 1.7 M NaCl) at least. Seawater, saline soils, salt lakes, brines, and alkaline saline habitats are considered as the best habitats for isolating halophilic Actinobacteria. Generally, most of the halophilic Actinobacteria have been isolated from saline soils. Halophilic Actinobacteria isolated from marine environments are assigned to a few genera, including Micromonospora, Rhodococcus, and Streptomyces. The other group includes Dietzia, Salinispora, Marinophilus, Solwaraspora, Salinibacterium, Aeromicrobium, Gordonia, Microbacterium, Mycobacterium, Nocardiopsis, Pseudonocardia, Actinomadura, Saccharopolyspora, Streptosporangium, Nonomuraea, Williamsia, and Verrucosispora.

Endophytic Actinobacteria

Endophytic Actinobacteria are defined as those that inhabit the internal part of plants, causing apparently no visible changes to their hosts. These Actinobacteria play specific roles, for instance, protecting the host plants against insects and diseases. Endophytic Actinobacteria constitute a large part of the rhizosphere, which are also found inside plants in which the extensively studied species are from the genus Frankia, nitrogen-fixing bacteria of nonleguminous plants, and a few species of the genus Streptomyces that are phytopathogens. Generally, the endophytic Actinobacteria include Streptomyces, but the genera Streptoverticillium, Nocardia, Micromonospora, Kitasatospora, Pseudonocardia, Microbispora, Kibdelosporangium, Actinopolyspora, Nocardioides, Brevibacterium, Actinomadura, Glycomyces Plantactinospora, Polymorphospora, Promicromonospora, and Streptosporangium are also found in the plants, such as Palicourea longifolia, Calycophyllum acreanum, Monstera spruceana, Croton lechleri, Cantua buxifolia, Siparuna crassifolia, and Eucharis cyaneosperma.

Symbiotic Actinobacteria

About 15% of the world's nitrogen is fixed naturally by the symbiotic relationships between various species of the Frankia belonging to the family of Actinobacteria. The plants that form symbiotic relationships with Frankia are called actinorhizal plants. Researchers have found over 160 plants that have Actinobacteria as their host, including alders, Russian olive, bayberry, sweet fern, bitterbrush, and cliff rose. The Frankia have the ability to provide most or all of the host plant's nitrogen needs. Numerous Frankia species including Casuarina isolates form nitrogen-reducing (NIR) vesicles in vitro and in planta. These nitrogen-fixing bacteria and their host plants are often pioneer species on young nitrogen-deficient and disturbed soils such as moraines, volcanic flows, and sand dunes.

Endosymbiontic Actinobacteria

An endosymbiont is any organism that lives within the body or cells of another organism. Endosymbiosis process is sometimes obligate, that is, either the endosymbiont or the host cannot survive without the other. Members of the phylum Actinobacteria have been identified as abundant members of sponge-associated microbial communities. Mycobacterium along with Micrococcus, Micromonospora, Microbacterium, Brevibacterium, Kocuria, Corynebacterium, Rhodococcus, Brachybacterium, Rubrobacter, Streptomyces, Dietzia, Salinispora, Actinokineospora, Gordonia, Arthrobacter, Nocardiopsis, and Rothia species were found to live as endosymbionts in marine sponges Callyspongia aff. Implexa, Aplysina aerophoba, Spheciospongia vagabunda, Hemimycale culumella, Hyrtios erecta, Dysidea tupha, Callyspongia sp., Dysidea avara, Amphimedon sp., and Negombata magnifica. However, the Actinobacterial endosymbionts have also been reported in other group of animals, such as Hylobates hoolock, Rhinopithecus roxellanae, Rhinopithecus bieti, Panthera tigris altaica, Panthera tigris tigris, Panthera tigris amoyensis, Ailurus fulgens, Cavnlvara zlrsidae, Ursus thibetanus, Cervus elaphus, Elaphurus davidianus, and Vicugna pacos.

Gut Actinobacteria

Though Actinobacteria are found in various diverse habitats, some are also known to form intimate associations with invertebrates and vertebrates. Symbiotic interactions are essential mainly for the survival and reproduction because they play a crucial role in nutrition, detoxification of certain compounds, growth performance, and protection against pathogenic bacteria. Many studies have shown that some symbiotic Actinobacterial species, that is probiotics, control bacterial diseases in livestock, poultry, and aquaculture. They also take part in host health by converting the feedstuffs into microbial biomass and fermentation end products that can be utilized by the animal host. Tan et al isolated Streptomyces, Nocardiopsis, and Oerskovia from healthy goat feces. Similarly, Latha and Dhanasekaran isolated 87 Actinobacterial cultures from different feces of goat and chicken collected from various locations in Pudukkottai and Tiruchirappalli Districts, Tamil Nadu, among which 45 isolates were selected for the screening of antibacterial activity and extracellular digestive enzyme production. The ability of the probiont Streptomyces sp. JD9 from gut of chicken possesses all the characteristics needed to satisfy the indigenous Actinobacterial probiont for enhanced broiler production.

Harmful Effects

Actinobacterial Plant Diseases

A number of significant plant diseases are caused by Actinobacteria. Actinobacteria currently assigned to the genus Corynebacterium cause a variety of diseases. Serodiagnosis has been used to detect and identify Corynebacterium insidiosum and Corynebacterium sependonicum, but most of these Corynebacteria are still identified by inoculation tests in host plants. Most phytopathogenic Corynebacteria are considered to produce their antagonistic effects by the production of hormones, polysaccharides, and toxins, whereas the potential of some strains to form biosurfactants may help in attachment to the host. The taxonomy of the phytopathogenic Corynebacteria is unsettled, though some prefer to retain these organisms in the genus Corynebacterium. Others accept that there is an overwhelming case for restricting Corynebacterium for animal pathogenic Corynebacteria and related strains and a consequent need to reclassify the plant pathogens. Corynebacterium betae, Corynebacterium oortii, and Corynebacterium poinsettiae are genetically identified with Corynebacterium flaccumfaciens and have many properties in common with the genus Curtobacterium, which would serve as a suitable niche for them. C. insidiosum and Corynebacterium michiganense, also fell into a single genospecies, were recovered in a loose DNA homology group with Corynebacterium nebraskense and Corynebacterium sepedonicum and together with the latter probably from the nucleus of a new genus.

Table: Plant diseases caused by Actinobacteria.

Arthrobacter ilicis	B light of holly (flex opaca).
C. betae	Wilt and leaf spot of red beet (Beta vulgaris).
C. flaccumfaciens	Wilt of bean (Phaseolus vulgaris).
C. insidiosum	Wilt and stunting of alfalfa (Medicago sativum).
C. michiganense	Canker of tomato (Lycopersicon esculentum) and some other solanaceous plants.
C. nebraskense	Wilt and blight of corn.
C. oortii	Spot of tulip leaves and bulbs.
Corynebacterium Poinselliae	Stem canker and leaf spot of poinsettia (Euphorbia pulcherrima).
Corynebacterium rathayi	Gumming of cereals.
C. sepedonicum	Wilt and tuber rot of potato (Solanum tuberosum).
Nocardia vaccinii	Galls and bud proliferation in blueberry plants (Vaccinium).
Rhodococcus fascians	Leaf gall in many plants, fasciation of sweet pea.
S. aureofaciens S. flaveolus. S. griseus	Common scab of potato.
S. ipomoeae	Sweet potato scab.
S. scabies, Streptomyces sp.	Common and russet scab of potatoes, sugar beet, etc.

Rhizobium

Rhizobium are a group of Gram-negative soil bacteria that are well known for their symbiotic relationship with various leguminous (soybeans, alfalfa etc).

There are different types of Rhizobium that are categorized on the basis of the rate of growth and the type of plant they are associated with.

Some species of Rhizobium include:

- R. leguminosarum.
- R. alamii.

- R. lentis.
- R. japonicum.
- R. metallidurans.
- R. smilacinae.
- R. phaseoli.
- R. trifolii.

Classification

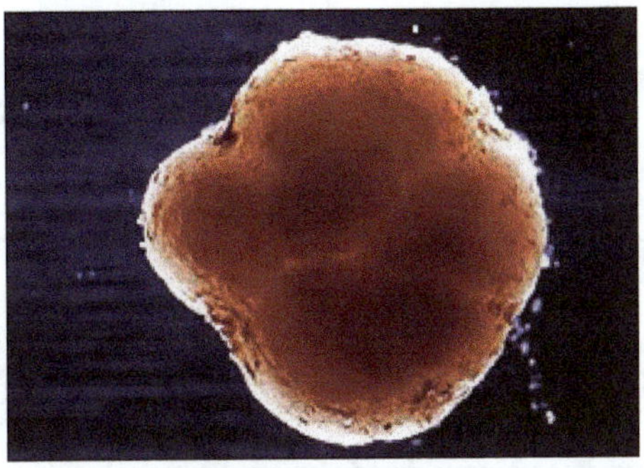

Rhizobium bacteria.

Along with Bradyrhizobium, Sinorhizobium, Mesorhizobium, Azorhizobium, and Allorhizobium, Rhizobium is a soil Rhizobia, which means that it consists of bacteria with the ability to fix nitrogen. As such, it presents a significant advantage to the plants it infects by contributing to their growth and development.

It is classified as follows:

- Kingdom (Bacteria): Like other bacteria, Rhizobium cells lack membrane-bound organelles. They also have ribosome, cytoplasm and flagella.
- Phylum (Proteobacteria): Phylum proteobacteria includes Gram-negative bacteria including those responsible for fixing nitrogen.
- Class (Alphaproteobacteria): Class Alphaproteobacteria is composed of a wide range of Gram-negative with varying life characteristics.
- Order (Rhizobiales): This is an order of Gram-negative Alphaproteobacteria that form a symbiotic relationship with plant roots (where they fix nitrogen.)
- Family (Rhizobiaceae): Along with such genera as Agrobacterium, genus Rhizobium makes up the family Rhizobiaceae. While some in this family have been shown to negatively affect plant development, Rhizobium (a member of Rhizo-bia) contribute to plant nutrition through nitrogen-fixation.

- Species: Includes such species as Rhizobium leguminosarum and Rhizobium lentil.

In addition to this classification, Rhizobium bacteria are also categorized based on the species of legume that they nodulate. This type of grouping is known as cross-inoculation.

Based on studies on a wide variety of legumes, it became evident that not all Rhizobia are capable of nodulating all types of legumes. This resulted in a need to group Rhizobium into appropriate groups (cross-inoculation groups) which are simply groups of legumes that given species of Rhizobium nodulate.

The cross-inoculation groups include:

- Clover groups: R. trifolii infects and nodulates plants of genus Trifolium (clovers/trefoil).
- Alfalfa groups: R. meliloti infects and nodulates the roots of medicago, melilotus and medicago.
- Bean group: R. phaseoli infects and nodulates plants of genus Phaseolus (e.g. beans).
- Lupine group: R. lupine nodulates lupines and serradella (Ornithopus).
- Pea group: R. leguminosarum infects and nodulates pea, sweet pea, lentil, and vetch.
- Soybean group: R. japonicum nodulates Glycine such as soybean.
- Cowpea group: Rhizobium sp. nodulates cowpea, pegionpea, lespedza, groundnut and kudz among a few others.

Characteristics (Rhizobium Leguminosarum)

Discovered and described in 1889, R. leguminosarum is the type species of Rhizobium in the same way Rhizobium has been the type genus of family Rhizobiaceae. As such, it can be viewed as a representative of the genus (Rhizobium).

Some of the characteristics of the bacteria include:

- They appear as elongated rods when viewed under the microscope.
- Like a number of other bacteria, Rhizobium leguminosarum do not form spores in their life cycle.
- They posses several flagella on their polar end. This allows them to move from one location to another.
- They are aerobic. As such, they need oxygen for respiratory purposes.

- There are various strains of the bacteria some of which have granules.
- They are Gram-negative bacteria.
- Although they can tolerate higher temperatures of about 38 degrees Celsius, Rhizobium. leguminosarum ideally grow in temperatures of between 20 and 28 degrees Celsius.
- Apart from various types of carbohydrates, the bacteria also uses nitrates and nitrite, ammonium salts and various amino acids among others for development.

There are different strains which include:

- Rhizobium leguminosarum biovar viciae 3841.
- Rhizobium leguminosarum USDA2370T.
- Rhizobium pisi DSM30132T.
- Rhizobium fabae CCBAU33202T.

Infection (Penetration)

Rhizobium species like R. leguminosarum can be found in soil. However, the root of leguminous plants (lentil, sweetpea etc) is their primary habitat. In the soil, various leguminous plants release various exudates (dicarboxylic acids etc) that attract Rhizobium species.

Flavanoids have also been shown to play an important role attracting the bacteria given that they are easily absorbed through the membrane of the organisms (passively). Once the bacteria detect these chemicals, they actively swim towards and attach to the legume root.

In addition to attracting the bacteria, these chemicals (flavanoids in particular) also play an important role of activating genes involved in producing Nod factors. Here, then, attraction to the legume roots is followed by transcription of Nod genes in preparation of the symbiotic relationship.

For the plant, Nod factors stimulate the branching of root hair, hydrolysis of the cell wall as well as deformation of the cell wall. Having attracted the bacteria through the exudates, the changes in the plant roots make it easy for the organism to enter the cells of the root hair for symbiosis.

When the bacteria comes into contact with the root hair, they cause the plasma membrane of the cells to invaginate. As the bacteria penetrates the cell, the plant produces new cell wall material at the site to not only cover the bacteria, but also allowing them to enter deeper into the root hairs.

Once the bacteria infects the cells of the root hair, the symbiotic process may produce the following types of nodules:

- Determinate nodules: Determinate are found in plants like soybeans, are spherical in shape with large lenticels. Compared to indeterminate nodules, determinate nodules are formed once the tip of the root hairs start growing. However, they lack the persistent meristem found in indeterminate nodules.
- Indeterminate nodules: Compared to determinate nodules, indeterminate nodules are cylindrical in shape and frequently branched. They are often found in plants like peas and alfalfa and start developing even before the growth of the root tip.

Whereas indeterminate nodules are found in temperate regions, determinate nodules are commonly formed in tropical and sub-tropical areas.

Determinate nodules produce ureide products while indeterminate nodules amide products.

Nitrogen Fixation

Legumes inoculated with bacteria of genus Rhizobium to induce nitrogen-fixing nodules in roots of legumes: Terraprima.

Nitrogen fixation in the nodules begins when the nodules fully mature. Here, it is worth noting that nitrogen fixation involves the conversion of atmospheric nitrogen into organic compounds (particularly ammonia) that can be used for plant development.

This process requires two important genes (nif and fix). These genes play an important role of producing several crucial enzymes that are involved in the nitrogen fixation.

The process requires the following:

- Enzymes (dinitrogenase and dinitrogenase reductase).
- ATP energy.
- Atmospheric Nitrogen (Nitrogen molecules).

During nitrogen fixation, the enzyme nitrogenase is involved in the breaking of the bonds that hold Nitrogen atoms together (covalent bonds). In their atmospheric state, nitrogen molecules are non-reactive given that they are bound by covalent bonds. By breaking this bond, the Nitrogen atoms are free to form bonds with other atoms.

This process (breaking down the covalent bonds) requires a lot of energy. Given that Rhizobium bacteria are not capable of making their own food for energy, they rely on the plant (in the rhizospere) to provide sources of energy.

By using energy sources from the plant, the bacteria gains sufficient energy that makes it possible for the enzymes to break down Nitrogen molecules into Nitrogen atoms.

Nitrogenase enzyme is oxygen sensitive. However, high metabolic activity of the bacteria as well as a diffusion barrier developed at the nodule periphery help protect the enzyme from the high level of oxygen.

Nitrogen fixation in the nodules takes place while the Nitrogen molecules are attached to the enzyme. While the nitrogen is bound to the enzyme, electrons provided by ferredoxin make it possible for the Iron protein of the enzyme to be reduced. This protein then binds to the ATP, which in turn causes molybdenum-iron protein (also a component of the enzyme) to be reduced.

This reduction provides electrons required to break down the Nitrogen molecules and produce a compound known as Diimide (NH_2). The process is repeated two more times which further reduces the Diimide into two ammonia molecules.

The process is represented as follows:

$$N_2 + 8H + 8e^- + 16Mg-ATP\ 2NH_3 + H_2 + 16MgADP + 16Pi$$

The relationship between leguminous plants and Rhizobium bacteria is referred to as a symbiotic relationship because the bacteria and the plant benefit each other. While the plant's rhizosphere provides shelter and a source of energy for the bacteria, the bacteria converts atmospheric nitrogen to ammonia which is required for proper growth and development of the plant.

Biofertilizer

Nitrogen is one of the most important elements in nature given that it is used to make various products that plants require for their development. For instance, using such products as ammonia and nitrates that are made from nitrogen, plants are able to form the protein they need for their development.

While various chemical fertilizers are successfully used to increase yields, they are also expensive and tend to pollute the environment. Given that Rhizobium bacteria have a good and beneficial relationship with various leguminous plants, there has been increased interest to use them as biofertilizers.

In small scale scenarios, the inoculation of Rhizobium in various cereal grains has been shown to help increase yields. This was also shown to result in an increased uptake of nitrogen thus benefiting the plants.

On a large scale, production of the inoculants begins with identifying the most effective strain of bacteria through host-specificity. Here, research helps identify the most effective strain of Rhizobium that is then cultured for large scale production.

During mass culture production, the bacteria is cultured in large flasks composed of such carbohydrates as mannitol, arabinose and sucrose among others. This helps produce fertilizer products that have been shown to help enhance photosynthesis and thus yields such plants as banana and rice among others significantly.

Biofertilizer presents a significant advantage in that they not only increase yields for farmers, but are also safe and do not negatively affect the soil and environment they are used in. Studies in different parts of the world such as Sudan have found this method to be particularly beneficial for a variety of crops.

Culture

Sample Preparation

- Carefully uproot a leguminous plant (sweetpeas, beans etc).
- Gently wash the roots using sterilized water to remove any soil, mud etc.
- Using a clean knife/blade, carefully remove/detach the nodules (which are visible to the naked eye) from the roots and place them in a container with clear water.
- After washing the nodules, place them in a container with a solution of merculic chloride for a few minutes - this helps disinfect and remove any microorganism on the surface of the nodule.
- Wash the nodules using alcohol and then wash with clean water several times.
- Place the nodules in a test-tube and add a few drops of water (distilled water).
- Using a glass rod, crush the nodules to release the bacteria into the water.

Culture Procedure

- Gently pour mannitol-agar medium into 3 sterile test-tubes.
- Add a few drops of the sample (rhizobium water) into the test-tubes using sterilized droppers.

- Mix the contents thoroughly.
- Pour the contents into different Petri dishes and incubate at 45 degrees Celsius.
- Turn the Petri dish when the medium solidifies.
- Allow the culture to incubate for about 3 days.

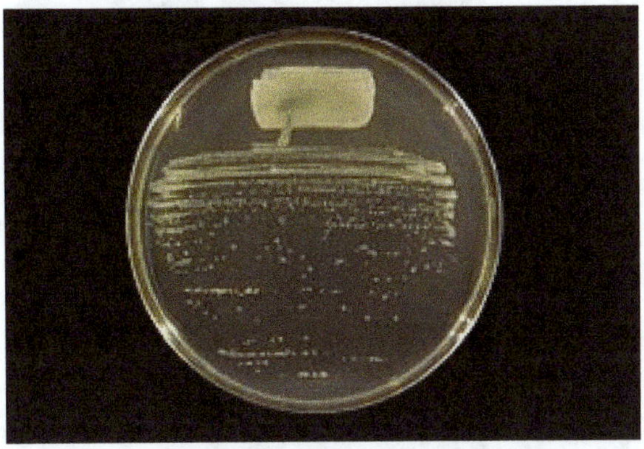

Rhizobium tropici strain BR816 streaked to single colonies
on Tryptone-Yeast Extract (TY) agar:Ninjatacoshell.

Gram Staining

Using a sample from the culture, Gram-staining involves the following steps:

- Fix the bacteria using methanol. Having placed the sample at the central part of a clean slide, add a few drops of methanol to fix. This helps preserve the morphology of the bacteria.
- Allow the slide to dry and pour the primary stain on the sample (crystal violet) - Allow to stand for a few minutes.
- Add a few drops of the mordant (Iodine solution).
- Decolorize the sample using ethanol/acetone gently.
- Pour safranin on the sample (counter-stain) for a few minutes.
- Wash gently with water to remove any excess counter-stain.
- Mount and view under the microscope.

Observation

The bacteria will appear as red/pink rods under the microscope with several flagella on one polar end.

When measured, they are about 0.5 to 0.9um in width and 1.2 to 3.0 um length.

Harmful Bacteria in Soil

Harmful bacteria found within soil can cause potential harm to humans, plants and trees. Some forms of bacteria can produce poisonous toxins, which can be fatal if the spores of such bacteria are inhaled, ingested or transferred through a wound.

There are various types of harmful bacteria found in soil.

Bacillus Species

There are a few variety of Bacillus. Bacillus cereus is a bacteria commonly found in soil. B. cereus is capable of withstanding extreme conditions, such as heat. Food grown in soil containing B. cereus can become susceptible to contamination. It is also possible to inhale aggravated B. cereus spores, or have spores enter broken skin when you don't wear gloves while gardening. According to the Textbook of Bacteriology, B. cereus contains three types of enterotoxins. Enterotoxins are toxins produced by bacteria and are responsible for causing the vomiting and diarrhea associated with food poisoning.

Food contaminated with B. cereus can result in food poisoning.

Crown Gall Disease

Crown gall affects plants and trees.

Agrobacterium tumefaciens is a form of bacteria that causes disease in plant tissue. If A. tumefaciens enters a healthy tree or plant through the root or stem from the soil, the bacterium will parasitize the tree or plant. The host of A. tumifaciens will succumb to tumor development and changes in plant metabolism. Tumors can begin as white callused tissue on the tree or plant. To prevent crown gall disease, it is important for plant life to be sustained outside of contaminated soil.

Anthrax Development

B. anthracis can survive in soil for years.

Anthrax is caused by the bacterium Bacillus anthracis. B. anthracis can survive for years within soil. When the bacterium produces spores, the potential for contamination becomes possible. Spores can be disturbed during gardening. According to the Directors of Health Promotion and Education, the inhalation of spores from contaminated soil can result in illness. Anthrax is also responsible for producing a toxin that can result in skin ulcers, respiratory distress, fever, vomiting, diarrhea, nausea and possible death.

Anaerobic Bacteria

C. perfringens is most commonly found in soil and water.

Anaerobic bacteria don't require oxygen to survive. One particular species, Clostridium perfringens can be found virtually everywhere. The bacteria can be found in the intestines of humans and animals. However, the bacteria is predominantly found in soil and water. C. perfringens is one of the most common bacterium responsible for food-borne illnesses. If C. perfringens is exposed to extreme temperatures, such as heat, the bacterium will sporulate, or form new spores. The new spores are extremely resistent; which can make outbreaks of illness possible.

Pathogenic Bacteria in Soil

Pathogens are microbes that cause disease in their hosts. While plants usually have more to fear from soilborne pathogenic fungi than from soilborne bacterial diseases, a variety of pathogenic bacteria that live in soil cause disease in plants as well. There are too many species of soil plant pathogens to list them all, but the following are some of the most important or common. A few species of bacteria in the soil can prove dangerous or even deadly to humans as well.

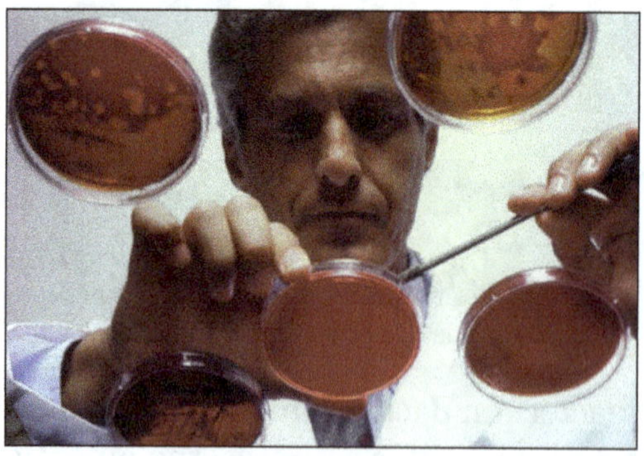

Clostridium Botulinum

C. botulinum is a gram-positive bacterium (i.e. a bacterium that tests positive on the Gram staining test) that forms hardy, durable spores. When the spores find the right conditions, they germinate and begin to grow. They can produce a lethal protein called botulinum toxin, which acts on the nervous system in humans and other mammals to block release of a key neurotransmitter called acetylcholine. C. botulinum spores are found in soil and water in many regions around the world. They are anaerobic and only grow in the absence of oxygen; acidic conditions inhibit their growth. Although C. botulinum sometimes causes fatal cases of a food-borne illness called botulism, the powerful toxin produced by the bacterium has also become useful for paralyzing muscle tissue in humans as part of various medical or cosmetic procedures. When used as a therapeutic agent, botulinum toxin is known by its trade name, Botox.

Erwinia Carotovora

E. carotovora is a highly destructive pathogen that causes disease in a wide range of plants, including potatoes and most other fleshy vegetables. This gram-negative, rod-shaped bacterium can invade the crops in the field or while in storage. The subspecies E. carotovora carotovora survives well in soil and surface waters; it can grow in the root zones of many other nonhost crop or even weed species, invading potatoes or other vulnerable crops when opportunity offers. Common potato diseases caused by E. carotovora include blackleg, aerial stem rot, and tuber soft rot.

Streptomyces Scabies

Streptomyces scabies are gram-positive bacteria that together with two other Streptomyces species (S. acidiscabies and S. turgidiscabies) can infect potatoes and other root crops like beets, radishes, turnips, carrots and parsnips. It's unusual in that as it grows, the cells form filaments similar to those formed by fungi — albeit far smaller. Filaments of S. scabies can then break off to form spores. The bacteria spread by invading young, fresh tissues and gaining entry to older tissue by way of wounds or natural openings.

Xanthomonas

Xanthomonas is a genus that includes many important plant pathogens. Although they do not persist for very long in the soil, some species can survive in the soil over the winter and infect a new crop in the spring. Like all other proteobacteria, they are gram-negative. The species X. campestris causes black rot in a wide variety of plants and is perhaps the most important member of the genus. Ironically, the same bacterium is grown commercially to produce xanthan gum, a common food additive.

Fungi

Fungi are microscopic cells that usually grow as long threads or strands called hyphae, which push their way between soil particles, roots, and rocks. Hyphae are usually only several thousandths of an inch (a few micrometers) in diameter. A single hyphae can span in length from a few cells to many yards. A few fungi, such as yeast, are single cells.

Hyphae sometimes group into masses called mycelium or thick, cord-like "rhizomorphs" that look like roots. Fungal fruiting structures (mushrooms) are made of hyphal strands, spores, and some special structures like gills on which spores form. A single individual fungus can include many fruiting bodies scattered across an area as large as a baseball diamond.

Fungi perform important services related to water dynamics, nutrient cycling, and disease suppression. Along with bacteria, fungi are important as decomposers in the soil food web. They convert hard-to-digest organic material into forms that other organisms can use. Fungal hyphae physically bind soil particles together, creating stable aggregates that help increase water infiltration and soil water holding capacity.

Soil fungi can be grouped into three general functional groups based on how they get their energy.

- Decomposers: Saprophytic fungi: convert dead organic material into fungal biomass, carbon dioxide (CO_2), and small molecules, such as organic acids. These fungi generally use complex substrates, such as the cellulose and lignin, in wood, and are essential in decomposing the carbon ring structures in some pollutants. A few fungi are called "sugar fungi" because they use the same simple substrates as do many bacteria. Like bacteria, fungi are important for immobilizing, or retaining, nutrients in the soil. In addition, many of the secondary metabolites of fungi are organic acids, so they help increase the accumulation of humic-acid rich organic matter that is resistant to degradation and may stay in the soil for hundreds of years.

- Mutualists: The mycorrhizal fungi: colonize plant roots. In exchange for carbon from the plant, mycorrhizal fungi help solubolize phosphorus and bring soil nutrients (phosphorus, nitrogen, micronutrients, and perhaps water) to the plant. One major group of mycorrhizae, the ectomycorrhizae, grow on the surface layers of the roots and are commonly associated with trees. The second major group of mycorrhizae are the endomycorrhizae that grow within the root cells and are commonly associated with grasses, row crops, vegetables, and shrubs. Arbuscular mycorrhizal (AM) fungi are a type of endomycorrhizal fungi. Ericoid mycorrhizal fungi can by either ecto- or endomycorrhizal.

- The third group of fungi, pathogens or parasites, cause reduced production or death when they colonize roots and other organisms. Root-pathogenic fungi, such as Verticillium, Pythium, and Rhizoctonia, cause major economic losses in agriculture each year. Many fungi help control diseases. For example, nematode-trapping fungi that parasitize disease-causing nematodes, and fungi that feed on insects may be useful as biocontrol agents.

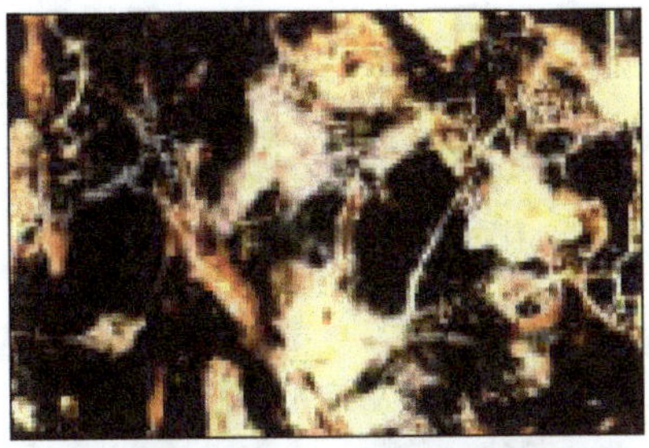

Mycorrhizal structure and fungal hyphae.

Many plants depend on fungi to help extract nutrients from the soil. Tree roots (brown) are connected to the symbiotic mycorrhizal structure (bright white) and fungal hyphae (thin white strands) radiating into the soil.

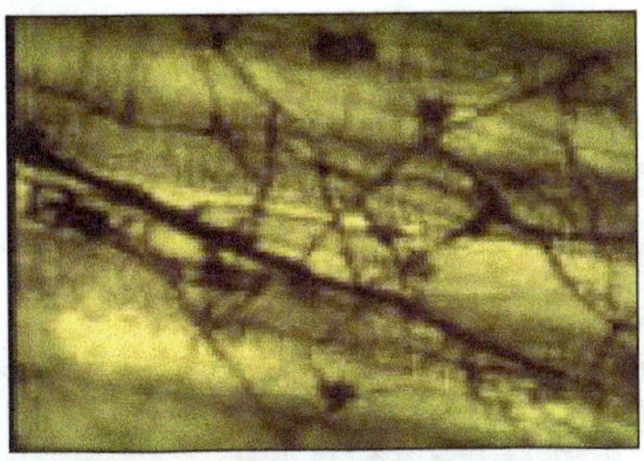

Fungus decomposing leaf veins.

Fungus beginning to decompose leaf veins in grass clippings.

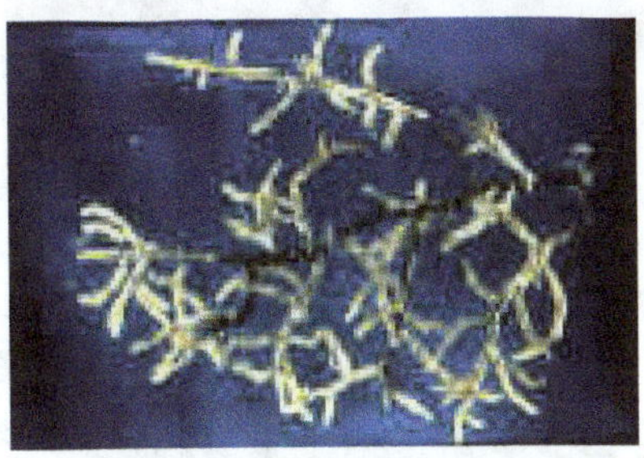

Ectomycorrhizae fungi.

Ectomycorrhizae are important for nutrient absorption by tree and grape roots. The fungus does not actually invade root cells but forms a sheath that penetrates between plant cells. The sheath in this photo is white, but they may be black, orange, pink, or yellow.

The dark, round masses inside the cells of this clover root are vesicules for the arbuscular mycorrhizal fungus (AM).

Saprophytic fungi are commonly active around woody plant residue. Fungal hyphae have advantages over bacteria in some soil environments. Under dry conditions, fungi can bridge gaps between pockets of moisture and continue to survive and grow, even when soil moisture is too low for most bacteria to be active. Fungi are able to use nitrogen up from the soil, allowing them to decompose surface residue which is often low in nitrogen.

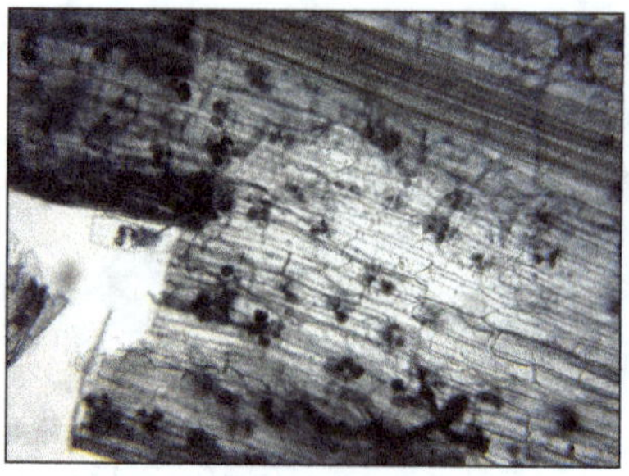

Arbuscular Mycorrhizae.

Fungi are aerobic organisms. Soil which becomes anaerobic for significant periods generally loses its fungal component. Anaerobic conditions often occur in waterlogged soil and in compacted soils.

Fungi are especially extensive in forested lands. Forests have been observed to increase in productivity as fungal biomass increases.

Arid rangeland.

In arid rangeland systems, such as southwestern deserts, fungi pipe scarce water and nutrients to plants.

Mushroom.

Mushrooms, common in forest systems, are the fruiting bodies made by a group of fungi called basidiomycetes. Mushrooms are "the tip of the iceberg" of an extensive network of underground hyphae.

Mycorrhizal Fungi in Agriculture

Mycorrhiza is a symbiotic association between fungi and plant roots and is unlike either fungi or roots alone. Most trees and agricultural crops depend on or benefit substantially from mycorrhizae. The exceptions are many members of the Cruciferae family (e.g., broccoli, mustard), and the Chenopodiaceae family (e.g. lambsquarters, spinach, beets), which do not form mycorrhizal associations. The level of dependency on mycorrhizae varies greatly among varieties of some crops, including wheat and corn.

Land management practices affect the formation of mycorrhizae. The number of mycorrhizal fungi in soil will decline in fallowed fields or in those planted to crops that do not form mycorrhizae. Frequent tillage may reduce mycorrhizal associations, and broad spectrum fungicides are toxic to mycorrhizal fungi. Very high levels of nitrogen or phosphorus fertilizer may reduce inoculation of roots. Some inoculums of mycorrhizal fungi are commercially available and can be added to the soil at planting time.

Mycorrhizal fungi link root to soil particles.

Mycorrhizal fungi link root cells to soil particles; sand grains are bound to a root by hyphae from endophytes (fungi similar to mycorrhizae), and by polysaccharides secreted by the plant and the fungi.

Role of Soil Fungus

Fungi are an important part of the microbial ecology. The majority of fungi decompose the lignin and the hard-to-digest soil organic matter, but some fungi consume simple sugars. Fungi dominate in low pH or slightly acidic soils where soils tend to be undisturbed. Fungi break down the organic residues so that many different types of microbes can start to decompose and process the residues into usable products.

Approximately 80 to 90 percent of all plants form symbiotic mycorrhizae fungi relationships by forming hyphae networks. The hyphae are about 1/60 the diameter of most plant root hairs and assist the plant in acquiring nitrogen, phosphorus, micronutrients and water in exchange for sugar produced by the plant. This mutually beneficial relationship is called a mycorrhizae network.

Hyphae interact with soil particles, roots, and rocks forming a filamentous body that promotes foraging for soil nutrients. These networks release enzymes into the soil and break down complex molecules that the filaments then reabsorb. Fungi act like natural recycling bins, reabsorbing and redistributing soil nutrients back to plant roots. Most hyphae are either pure white or yellow and are often misidentified as plant hair roots.

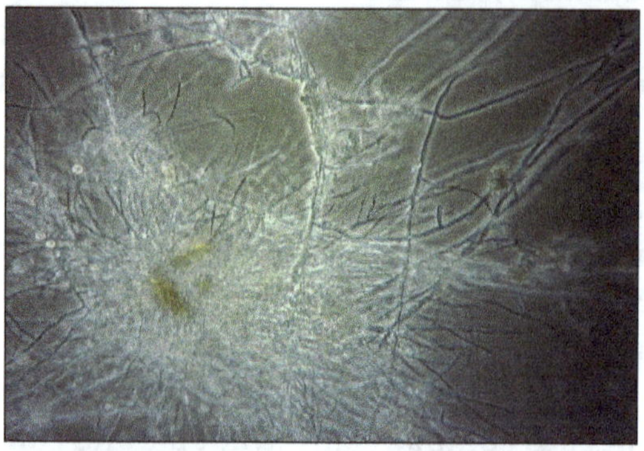

Soil fungus form mycorrhizae networks like a spider web to explore the soil profile for soil nutrients.

Mycorrhizal fungi are usually white or yellow while the root at the top is a light brown or tan color.

The rhizosphere is an area next to the root dominated by soil microbes where many chemical and biochemical process occur. Soil fungi make up 10 to 30 percent of the soil rhizosphere. Generally there are fewer individual fungi than bacteria but fungi dominate the total biomass due to their larger size in a healthy soil. Fungi biomass in the soil ranges from the equivalent of two to six cows in a healthy soil or 1,100 to 11,000 pounds of biomass.

Fungi prefer slightly acidic conditions, low disturbance soils, perennial plants, internal nutrient sources directly from the plant, and highly stable forms of organic residues with high carbon to nitrogen (C:N) values and slower recycling time. Bacteria dominate in highly disturbed ecosystems with fast nutrient recycling, low C:N values, prefer annual plants, and external nutrient additions outside the plant. Bacteria are single-celled organisms and need a film of water to survive, while fungi are multi-celled organisms that grow rapidly and in great lengths in the soil (feet or meters). This allows fungi to bridge gaps in the soil so as to transport nutrients relatively far distances back to the plants.

Taxonomy and Functional Groups of Fungi

There are at least 70,000 different species of fungi identified but it is estimated that there may be 1.5 million species worldwide. Genetically, fungi evolved a billion years ago and are closely related to plants and animals. Membrane bound organelles present in each cell are similar to those found in insects, plants and animals. Fungi have 80 percent or more of the same genes as humans.

There are four major groups of soil fungus: Zygomycota, Ascomycota, Basidiomycota, and Deuteromycota. Zygomycota are less than 1,000 species and are mostly common bread molds. Ascomycetes have about 30,000 species and are mostly yeasts used in baking. Basidiomycetes include most mushrooms, toadstools and puffballs, while Deuteromycota include the lichens and the mycorrhizal fungus. Fungi are classified as heterotrophs so the carbon source originates from the decomposition of organic compounds or residues.

Decomposers are also called saprophytic fungi which decompose cellulose and lignin in the soil. Sugar fungi called Zygomycetes decompose the simple sugars but most fungi decompose the more recalcitrant or hard-to-decompose organic residues high in cellulose, hemicellulose, lignin or cell walls. Some of the byproducts of this decomposition may turn to humus and remain in the soil for thousands of years.

Pathogenic fungi cause many agricultural root diseases including Phytophthora, Rhizoctonia, Phythium, and Verticullium, and downy mildew Ascomycetes fungi are microscopic in size and dominate in agricultural soils and grassland while the Basidiomycetes have large fruiting bodies or mushrooms that dominate in high residue and forested soil. Some fungi help to control diseases and predators including a nematode trapping fungi that feeds on insects and can be used as biological controls.

Mutualistic mycorrhizal fungi form a beneficial relationship with plants. Ingham states that "mycorrhizae grow within the root cells and are commonly associated with grasses, row crops, vegetables, and shrubs. Some plant species like the Cruciferae family (e.g., cabbage, broccoli, mustard and canola) and the Chenopodiaceae family (e.g., lambsquarters, spinach, beets, and oilseed radish) do not form mycorrhizae associations."

Ecological Plant-microbe Interactions

The microbes and plants together regulate many soil processes including the carbon cycle and nutrient recycling. The microbial species diversity and the total microbial population determine the ability of plants to obtain soil nutrients like nitrogen, phosphorus and micronutrients. Plant diversity and abundance may change the entire soil ecosystem through the release of root exudates that attract or inhibit the growth of specific organisms. These carbon-rich substances can range from less than 10 percent to as much as 20 percent of a plant's total carbon production. Plants secrete large amounts of specific carbon compounds into the soil to be used by the microbes as a food and energy source and to enhance and improve soil structure. By encouraging certain microbial species to grow, the microbes in return supply nutrients to the plant roots and encourage the microbes to protect the plant from pathogenic microbes. Plants feed, raise and encourage certain microbes just like farmers raise and feed plants and livestock for food and fiber.

Many plants have a preferred fungus to bacteria ratio (F:B ratio). Most vegetable crops prefer more bacteria with a F:B ratio of 0.3 to 0.8 F:B which is optimal for carrots, lettuce, broccoli and cole crops. Tomatoes, corn and wheat like an F:B ratio of 0.8 to 1:1. Lawns prefer a F:B ratio of 0.5 to 1:1 and on forested soils, trees grow better with a F:B ratio of 10:1 where the soil is more acidic. Many conventionally tilled agricultural soils have a F:B ratio of 0.1 to 0.3 with soils that are high in nitrogen, low carbon, neutral pH, and with disturbed soil conditions which promotes weed production. Annual crops prefer lower F:B ratios and perennials prefer a higher F:B ratio. Many plants cultivate certain species of both bacteria and fungus to increase nutrient extraction from the soil.

Fungi benefit most plants by suppressing plant root diseases and fungi promote healthier plants by attacking plant pathogens with fungal enzymes. Fungi also use antagonism to reduce competition by producing antibodies, which suppress other microorganisms from growing. They produce many vitamins which promote plant growth. By competing with other fungus for nutrients, beneficial fungi prevent pathogenic and disease-causing organisms from getting established.

Beneficial fungi along with some bacteria may also form protective webs and nets around roots and even leaves to protect the host plant. The fungus Trichoderma protects plant roots from attack by harmful microorganisms. Fungi also protect plants by supplying a protective sheath to supply both water and phosphorus to the plant roots during droughts.

Hyphae must be in close contact with living or dead organic soil residues to absorb nutrients, so they usually grow in association with other soil microorganisms. They compete aggressively for scarce nutrients, and competition usually results in a succession or change in microbial populations as nutrients are absorbed or depleted. Initial colonizers absorb simple sugars, amino acids, and vitamins from plant parts such as fruits, seeds and vegetables, and are classified as "sugar fungus." The dominance of these fungi is short-lived because waste products accumulate quickly.

Cellulose degraders appear next and they are the most diverse fungi and competitive. For example, degradation of straw with high C:N ratio (80:1) requires that fungus parasitize or decompose other fungi to obtain nitrogen for growth and enzyme production. Degradation of lignin is stimulated by low nitrogen. Lignin makes up 60 percent of the total mass of humus, but the low number of fungal species that can degrade lignin reduces competition.

Fungi generally reproduce asexually by spores (microscopic parts similar to plant seeds). Spore dispersal occurs in a variety of ways including triggers, jet propulsion, springs, and scents depending on the environment. Fungi spores may live 50 years in an inactive vegetative state as a spore. This allows the fungi to survive and remain viable by staying inactive until environmental conditions improve. Microbes spend most of their time hungry and looking for nutrient resources just to survive but fungus may remain viable for years. Most fungi survive by continual growth and translocation of nutrients in wide networks as they search for food or they survive through the production of resistant spores during times of stress. Natural soils tend to suppress germinating fungal spores, especially when nutrients are limiting.

The longevity of fungi has not been measured in many species but their open-ended growth suggests that they have a longevity measured in millions of years, because they are basically the same organism. For example, fairy fungal rings grow in ever widening circles, much like rings on a tree, and are measured in decades and centuries instead of days and weeks for most microbes.

Algae

Soil algae (both prokaryotes and eukaryotes) luxuriantly grow where adequate amount of moisture and light are present. They play a variety of roles in soil. One of the important role of blue-green algae is that it has revolutionised the field of agriculture microbiology due to use of cyanobacterial biofertilizer.

On the other hand they can also be used in reclamation of sodic soil i.e. alkaline soil, sewage treatment, etc. The prominent genera are Anabaena, Calothrix, Oscillatoria, Aulosira, Nostoc, Scytonema, Tolypothrix, etc.

Algae are important components of arid and semi-arid ecosystems. Furthermore, their distribution and condition may indicate the health of the environment. Additionally, the presence of algae leads to reduced erosion by regulating the water flow into soils. Similarly, they play a role in soil fertility, soil reclamation, bio-controlling of agricultural pests, and formation of microbiological crust, agricultural wastewater treatment and recycling of treated water. Human civilization depends on agriculture for its existence. The success of agriculture greatly depends on the fertility level of the soil. Like other organisms, algae which are found in different soil types, may help the soil to improve its characteristics such as, carbon content, texture, aeration and also nitrogen fixation. The magnitude of these improvements is greatly dependent on the physical and chemical characteristics of the soil, affecting the composition of the algal population.

Marine algae are used as fertilizers on farmlands close to the sea, examples include the large brown and red algae used as organic fertilizers; which are usually richer in potassium but poorer in nitrogen and phosphorus. The weed is usually applied direct and ploughed in, both as solid (processed into a seaweed meal) and as liquid fertilizer (concentrated extract of seaweeds). Probably the widest use of seaweeds in agriculture is as liquid fertilizers. The positive effect of liquid fertilizers is mainly explained by the high content of trace elements and growth regulatory substances (particularly cytokinins). Calcareous red algae known as maerl, are used in the United Kingdom and France to reduce soil acidity. Man's uses of algae, particularly marine algae, are far more diverse and economically important than generally realized. They are used as human food, in agriculture (fertilizer, manure, fodder and aquaculture), medicine, textile, paper and paint industries, chemical extracts from larger marine algae (example alginic acid, carrageenan or agar) are used in the manufacture of food industry, and diatomaceous earth (deposits of diatom frustules) is widely used as filtration and polishing materials. Algae are also important surface-binding agents which reduce erosion and can be used for wastewater treatment.

Cyanobacteria are a diverse group of prokaryotes. A common feature is their oxygenic photosynthesis, which is similar to that in algae and higher plants. As sunlight is their energy source and water, they generate oxygen in the light. Energy and reductants generated by photo-synthesis are usually used for carbon dioxide reduction. These microorganisms are distributed worldwide and improve the growth and development of the plants, with which they share the habitat, because they: 1) Contribute to soil fertility in many ecosystems; 2) Produce various biologically active substances and 3) Have higher efficiency in biosorption of heavy metals (bioremediation).

Improvement of Soil Fertility

Some cyanobacteria are able to reduce atmospheric nitrogen to ammonia, a process where oxygen evolved by photosynthetic activity in the same cell is detrimental to nitrogen fixation. Strategies to avoid oxygen range from temporal separation of nitrogen fixation and oxygen evolution (in many unicellular and filamentous, non heterocysts strains) to spatial separation and cellular differentiation into nitrogen fixing heterocysts (in filamentous cyanobacteria). Heterocysts are terminally differentiated cells whose interior becomes anaerobic, mainly as a consequence of respiration, allowing the oxygen-sensitive process of nitrogen fixation to continue. The regulation of dinitrogen fixation has been extensively studied in the heterocyst system.

Diazotrophic cyanobacteria require sunlight as a sole energy source for the fixation of carbon and nitrogen. Therefore, they have great potential as biofertilizers, and their use will decrease fuel demand for fertilizer pro-duction. The agronomic potential of heterocystous cyanobacteria, either free-living or in symbiotic association with water fern Azolla, has long been recognized. This had led to the development of small scale biotechnology involving the use of paddy soils with appropriate cyanobacterial strains as biofertilizers in rice fields, as has been reported in China, Egypt, Philippines and India.

Cyanobacteria are congenial biofertilizers for rice based cropping systems, being the major components of wetland rice ecosystems which are easily available and serve as the cheapest sources of natural biofertilizers. Whereas the in-corporation of genes into rice plants by using tissue culture and modern genetic tools remains as an ambitious research goal, the use of cyanobacterial diazotrophic technology in rice agriculture offers an immediate and even long term alternative to synthetic nitrogen fertilizers, particularly in developing countries and the world as a whole. However, one of the weak-nesses in the technology is the heavy application of several toxic agrochemicals, especially herbicides, which are reported in most cases as inhibitors of cyanobacterial diazotrophic growth, and in some cases as mutagenic. Therefore, a successful biotechnology requires the selection of suitable diazotrophic strains as biofertilizers that could tolerate the field-dose concentrations of herbicides.

Uptake of P and N

Cyanobacteria also have some soil phosphate-solubilizing species. Phosphorus (P) is the second important nutrient after nitrogen for plants and microorganisms. Most aquatic systems are resource-limited, where P and N are often the primary limiting nutrients. To ensure survival, a competitor must be able to maintain net population growth at resource levels less than those required by other species. Algae are particularly adapted to scavenge their environments for resources through structural changes, storage or increased resource utilization efficiency. Internal adjustments by algae involve biochemical and physiological adaptations, whilst they can also excrete substances to enhance nutrient availability. Algae excrete extracellular phosphatases almost immediately upon the onset of P limited conditions. Algae can also excrete other compounds and change the pH of their surroundings, which in turn can render adsorbed P available. In addition, algae can store resources like P in excess of their immediate needs. This excess or "luxury" uptake is clearly distinct from the Michaelis-Menten nutrient uptake kinetics which are based on external resource concentrations. Epply and Strickland concluded that the growth rate of phytoplankton is more closely related to the cellular nutrient content than to external concentrations. It is, therefore, necessary to establish a relationship between the cell quota of a nutrient and the growth rate of an alga. Such a relationship was given by Droop.

Cyanobacterial fertilization has been compared to inorganic fertilization on rice and lettuce seedlings. Biofertilizers are likely to assume greater significance as complement and supplement to chemical fertilizers in improving the nutrient supplies to cereal crops because of high nutrient turn-over in cereal production system, exorbitant cost of fertilizers and greater consciousness on environmental protection.

Nitrogen Fixation

Algae, especially cyanobacteria, may be the most important nitrogen-fixing agents in many agricultural soils. Their importance as nitrogen fixers in rice fields have been studied by several investigators. The great majority of cyanobacteria that fix nitrogen are probably hetero-cystous however, non-heterocystous cyanobacteria fix nitrogen as well. The nitrogen fixed by the algae is liberated and then re-assimilated by the higher plants. A large variety of cyanobacterial species are known to be nitrogen fixing and their importance in improving soil fertility for sustainable agriculture in submerged and irrigated rice cultivation is well recognized. The use of cyanobacteria as a biofertilizer for rice fields is very promising but limited due to fluctuation in quality and quantity of inoculum and its physiological attributes in varied agroecological regions. Utilization efficiency of fixed nitrogen by rice plants is often low and efforts are therefore being extended to isolate suitable strains of cyanobacteria that would be prolific not only in fixing atmospheric nitrogen but also in excreting it continuously, thus making it available to the growing rice plants.

Cyanobacteria are widely used in rice fields throughout Asia, where their enhancement of soil fertility by means of biological nitrogen fixation (so called algalization) in place of N-rich fertilizers, but their beneficial effects is not limited to that. The cyano-bacterium Tolypothrix tenuis is grown in cultures and added to rice fields. Aboul-Fadle et al. reported that inoculation at about 250 g dry mass ha-1 of the same species resulted in a 19.5% increase in rice yield as compared with 16.6% produced by a dressing of 25 kg ha-1 of ammonium sulfate. In Japan, rice fields are fertilized by water fern Azolla which multiplies rapidly and contains the symbiotic blue-green alga Anabaena which fixes gaseous N_2.

Source of Organic Matter

Algae are also important source of organic matter in soil. The organic matter formed from the death and decay of algae may get mixed in the soil and mucilage acts as binding agent for soil texture, thereby increasing the humus content and making it more habitable for other plants after some years. Humus accumu-lation is also important for moisture retention. In the U.S.S.R., filamentous forms of the Cyanophyceae, especially Oscillatoria, Schizothrix and Plectonema were found to be important in soil formation. In most cases, it is generally accepted that the incorporation of organic carbon via photosynthesis and of organic nitrogen via nitrogen fixation is the most important contributions of algae added to the soil. They also act as a reserve of inorganic nutrients.

Soil Reclamation

The difficulties in soil reclamation in arid and semi-arid regions are mostly the salinity conditions of large soil areas. Several studies have been carried out on the effect of salinity on the growth, metabolism and yield of the plants and algae. Some growth regulators such gibberellic acid (GA3) were used for improving the salt tolerance of the plants. From an economic point of view, growth regulators are expensive and are non-practical especially, when applied in large amounts. Algae play an economic role in soil reclamation increases soil fertility and improve the plant conditions under certain environmental factors.

Production of Extracellular Substances

Cyanobacteria excrete a great number of substances that influence plant growth and development. These micro-organisms have been reported to benefit plants by producing growth-promoting regulators (the nature of which is said to resemble gibberellin and auxin), vitamins, amino acids, polypeptides, antibacterial and antifungal substances that exert phytopathogen biocontrol and polymers, especially exopolysaccharides, that improve Plant growth substances.

While working on the algae of Indian paddy fields, Gupta and Lata observed that cyanobacteria accelerated seed germination and promoted seedling growth. In addition, they also observed that both the yield and the quality of the grains were improved in proteins content. It seems very likely that the beneficial effect of the algae on the rice crop may not be restricted to their capacity to fix atmospheric nitrogen alone, but also they have additional beneficial roles, such as releasing of bioactive substances. Mechanisms used by microbes to stimulate plant growth include biofertilization (increasing the supply of mineral nutrients to the plant), biological control (elimination of the plant enemies including microbial pathogens, insects and weeds) and direct plant growth production by delivering plant growth hormones. Biofertilization techniques using cyanobacteria are recommended for increasing the rate of seed germination and growth parameters of many plants.

Although microalgae and cyanobacteria are primary microbial photosynthetic agents of the soil, their ecological role is still not fully defined. However, it is obvious that some of their advantageous properties and beneficial effects influence plant/soil-systems. Two important potential uses of soil microalgae in crop production are as biofertilizers or soil conditioners. Recently, there is increasing interest in their antimicrobial- and PGR-compounds (Plant Growth Regulator). The effect of a great varieties of extracellular substances production by algae including cyanobacteria, play a valuable role in water habitats, as well as having a valuable role in enhancing the growth and germination of higher plants. For this reason, cyanobacteria algae were recommended as a biofertilizer due to the liberation of a very large portion of the bioactive substances from their assimilating nitrogen outside their cells. The previous inductions attributed to nitrogenase and nitrate reductase activities of cyanobacteria are associated with the surface of plants or the amino acids and peptides produced in algal filtrate and other compounds that stimulate growth of crop plants.

Microalgae are a biochemically diverse assemblage of microorganisms amenable to formation and mass culture. Including the cyanobacteria and nearly a dozen of eukaryotic classes, microalgae produce a wide array of compounds with biological activity. These may be nitrogenous, amino acid, vitamin B12 and biotin. Moreover, cyanobacteria have ability to exude also plant growth hormones including auxins like substances, cytokinin-like substances gibberellins or gibberellic-like substances, antibiotics, algicide, toxins, organic acids and pharmaceutically active compounds. A gibberellin-like substance has been isolated from the cyanobacterium Phormidium foveolarum and this is active in GA-bioassays. Moreover, chromatography identification of an excreted substance from Nostoc muscorum isolated from Argentine paddy fields revealed auxinic activity and characteristics similar to indole acetic acid. Growth-promoting substances were also detected by the effect of extracts of N. muscorum on seedlings of Panicum miliaceum. The height of millet plants as well as their dry weight was also increased by all the extracts. Information about cyanobacterial biomass or their substances being incorporated to other plants different to rice is scarce.

Bently in studying growth regulator production by phytoplankton showed that some strains of Anabaena and Oscillatoria exuded auxin-like substances. This was confirmed by the work of Likhitkar and Tarar who found that presoaked cotton seeds in different concentrations of exudate of N. muscorum increased germination rate, total length of seedlings and radicals. Adam revealed a stimulation effect of the cyanobacterium Nostoc muscorum on seed germination of wheat, sorghum, maize and lentil. He also concluded that the germination of seeds of the tested crop plants either in live cyanobacteria inoculum algal filtrate (exogenous) or boiled algal extract (endogenous) was significantly increased. He attributed it to the nitrogenase and nitrate reductase of the alga, or the amino acids and peptides produced in the algal filtrate and other compounds that stimulate growth of crop plants.

Moreover, Mahmoud and Amara revealed that biofertilizers enhanced the growth parameters, fruit yield as well as CO_2 evolution of tomato plant. Germination and related processes in wheat, sorghum, maize and lentil growth parameters and contents of nitrogenous compounds also increased. These reports indicated that the usage of biofertilizers enhances the growth parameters, fruit yield and vitamin C (ascorbic acid) in tomato plants. On the other hand, Tantawy and Musa studied the effect of cyanobacterial filtrates of Nostoc calcicola and Anabaena flos aquae on the seed germination and plant growth of some wheat, soybean and clover crop cultivars. They found that soaking the seeds for these crops in the filtrates of both cyano-bacterial strains increased the germination percentage either in the control treatment (water) and Watanabe medium. In this respect, Aref reported that soaking of rice seeds cultivar Sakha 102 in Nostoc sp. live filtrate had stimulated the seed germination percent age, reaching 90% of germination when compared to water soaked seeds. She also showed similar stimulation effect with both N. muscorum and Anabaena sp. (86.8 and 83.3, respectively). Additionally, Mohamed came to a conclusion that the treatment of rice seedlings with the cyanobacteria filtrates of Anabaena oryzae, N. calcicola, Microchaete tenera or Cylindrospermum muscicola had increased both shoots and roots lengths than those treated with water only. He owed this to the growth promoting like-substances secreted by the cyanobacterial strains in their filtrates.

Furthermore, Saffan and Mohamed studied the beneficial role of some bioactive substances released by cyanobacteria on the rate of germination Senna seeds as well as evaluation of the metabolic changes in medicinal plant Senna alexandrina. They reported that the exudates of Nostoc piscinale and N. muscorum increased up the rate of germination of Senna seeds, reaching 100 and 90% respectively after 60 h. Also, they found that the cyanobacterial exudates contained variable concen-trations of abscisic acid (ABA), gibberellic acid (GA3) and indole acetic acid (IAA) and other metabolites that might be implicated as allelochemical agents. They had also found out a significantly increased proteins and total soluble sugars in treatments with algal exudates, especially those of N. piscinale and N. muscorum. The allelopathic effects of cyanobacterial exudates (N. muscorum, N. piscinale and Anabaena fertilissima) on some biochemical constituents of cardon (Cynara cardunculus) have also been studied. The quantitative analysis of cyanobacterial exudates revealed the presence of phytohormones, amino acids, total soluble nitrogen and total reducing sugars. In addition, treatment with algal exudates stimulated the germination rate of cardon seeds after 96 h. Furthermore, the data revealed a significant increase in the total soluble sugar and protein contents in germinated seeds treated with different algae exudates.

Phytopathogen Biocontrol

Application of Chemical Pesticides

Soil is a dynamic system in which the physical, chemical and biotic components are in a state of equilibrium. Application of insecticides without considering the other soil constituents, disturb this equilibrium which adversely affects the productivity of the soil. Maintenance of the soil biota other than the harmful pests helps in better crop nutrient management and maintenance of soil health. Insecticides frequently exert inhibitory or stimulatory effects on the growth or other activities of micro-organisms, either in pure culture or in the field. Few works on pesticides distributions, types, toxicity, mechanism of actions, degradations, their tolerance by the organisms and other physiological processes were reviewed and summarized. Ghosh and Saha suggested that some pesticides actually are highly phytotoxic, such as carbaryl. The toxicity of carbaryl was studied also by Peterson et al. Ibraheem found that Larvin and Sevin when applied in Egyptian soils have very toxic effects on nitrogen fixing cyanobacteria Anabaena subtropica and A. variabilis. Blue-green algae, especially the nitrogen-fixers cyanobacteria, represent the major microorganisms which contribute soil fertility. These organisms play an important role in this system by providing a steady input of fixed nitrogen and other beneficial roles as previously discussed. Most of the soil and aquatic microscopic algae are sensitive to insecticides due to the fact that algae are engaged in photosynthesis and that many insecticides interfere with the process.

Until now many pesticides from different chemical and artificial sources were used as acaricidal, fungicidal and insecticide agents. These pesticides affect the distribution of fauna in their natural ecosystems. Also, the unjustifiable and unsafe applications of these pesticides on soils and plants cause accumulation of different undesirable chemicals in the crop, which may be a bigger un-direct factor in human diseases.

Biological Control

In the last decades, different researchers have studied the replacement of chemical pesticides by natural components of different plant and microalgal sources as insecticide agents, acaricide agents and fungicidal agents. These natural materials in addition to their lethal activities on pests, preserves the environment of pollution, maintain the equal distribution of fauna and also to keep the beneficial animals. Fungi and bacteria are the main biological agents that have been studied for the control of plant pathogens, particularly soil-borne fungi.

Cyanobacteria have received little attention as potential biocontrol agents of plant diseases. Caire et al. reported that different concentrations of dilute aqueous extract from nitrogen-fixing cyano-bacterium N. muscorum Ag. were efficient in the control of a damping-off. Also, it was found that the growth of the plant pathogens Sclerotinia sclerotiorum and Rhizoctonia solani, damping-off causal agents, was inhibited by extracts from cells of N. muscorum or by extracellular products of this cyanobacterium. With additional research, it should be possible to develop thin film formulations (polymers) of bactericidal and fungicidal cyanobacterial products that would confer protection against soil-borne pathogens that attack seeds and seedlings, when applied to high volume seeds and remain competitive. One of the modern and advanced biotechnological researches is that conducted on the using of different algal taxa of different habitats (marine, fresh and soil) as a biological control for many animal or plant diseases and also against agricultural pests. Some of these researches studied the anti-microbial activities against some human pathogenic bacteria, fungi and toxic micro-algae.

Others were conducted on the study of toxic effects of some algal metabolites against insects. For example, Mulla et al. found that a free floating unicellular Chlorella ellipsidea produces some substances which affect the development and immature stages of mosquitoes. Similar results were obtained by Nassar et al. who found that some cyanobacteria and green algae produce substances that inhibited larval development and delayed the survival and development of the adult female of mosquitoes. Van-Der Westhwzen and Eloff noticed that toxicity of Microcystis aeruginosa against the larvae of Culex pipiens increased gradually during the first days of exponential phase to the maximum and then gradually decreased at the beginning of the stationary phase.

Further studies by other workers were conducted on the possibility of prevention of carcinogenesis by some algal products. There is much evidence that the production of extracellular substances by blue-green algae is widespread and sometimes quantitatively important. The enhanced research activity on the subject of biological control is in line with increased effort and determination by microbiologists to adapt to the conceptual scheme of integrated pest management as an acceptable eco-system approach to disease control and to realize that biological control must become one of the basic components in pest management practices.

Exopolysaccharide

Cyanobacteria produce extracellular polymers of diverse chemical composition, especially exopolysaccharides that enhance microbial growth and as consequence, improve soil structure and exoenzyme activity. Maintenance of adequate levels of soil organic matter is essential for a sustainable and high production of crops. Cultivation alters the structural stability of soil and reduces the amount of N and soil organic matter.

The nature of this labile organic matter is not fully known, but a major portion of it could be microbial biomass. Reducing the amount of organic matter affects the stability of soil aggregates.

Incorporation of organic materials in soil promotes microbial growth and enzymatic activity in the soil. Some cyanobacteria excrete slime or mucilage that becomes dispersed around the organism and, to an extent, partially dissolves in the culture medium or in the soil solution. One way to positively affect nutrient content and soil structure is to add cyanobacteria.

Application of algal biofertilizers is also useful for the reclamation of marginal soils such as saline-alkali and calcareous soils. N. muscorum can improve the aggregate stability of a saline soil, where the increase in soil aggregation is mainly due to exopolysaccharide secreted by microorganisms or exopolysaccharide added to soil after death and cellular lysis. Cyanobacteria can be incor-porated into soil as organic matter and also as a source of enzymes as they produce acid and alkaline extracellular phosphatases that are active in solution or located in the periplasmatic space of the cell wall. Both biomass and exopolysaccharides incorporated into soil induce a growth promotion of other microorganisms and increased the activity of soil enzymes that participate in the liberation of nutrients required by plants.

Protozoa

Protozoa are non-cellular organisms which distinguishes them from the adult forms of all other animals, and most are microscopic in size. Free-living protozoa are found in every variety of habitat. Many such as Dinoflagellates, Foraminifera, and Radiolaria are important elements of marine plankton and after death their shells form the oozes of the ocean floor and subsequently the fossil beds of limestone deposits. In fresh water, Rhizopods, Flagellates, and Ciliates form the protozoan population and favourable conditions will result in a thick bloom of these forms. Protozoa are also found in hot-springs, salt-lakes, sewage filters and in the soil.

Distribution of Protozoa

Free-living protozoa in an encysted state are able to withstand the most adverse conditions for long periods. This fact combined with their microscopic size and the many avenues of transport which are open to them ensure that there can be no effective geographical barrier against their distribution. This is borne out by the numerous species which are found from the tropics to Greenland and from Europe to the Antarctic. So, when protozoan species are absent in an area it is usually attributable to unfavourable habitats.

Ciliates are found in both soil and fresh water but there are only a few species common to both habitats. In New Zealand, for example, more than 100 species of ciliates are known from soil and fresh water but only five of these occur in both habitats. In cases where a genus of ciliates is common to both environments the soil species are usually distinct from the freshwater species. Generally speaking, soil species are smaller than freshwater species and their small size is doubtless related to the restricted space in which most soil protozoa live. The different species of soil protozoa in New Zealand occur in characteristic groupings which can be correlated with soil type. Such a correlation excludes the possibility of explaining the distribution by random scattering. In soils of closely related types the species of the different systematic groups; amoebae, testaceans, holotrichous and spirotrichous ciliates show a common pattern or ratio although the actual species involved are not necessarily the same.

By using this pattern or ratio as a basis it is possible to compare soils under similar conditions in different parts of the world for although the species involved may vary from one geographical area to the next, the pattern will remain the same. Thus, the faunal patterns of samples taken under Podocarp vegetation in New Zealand and South Africa are very similar although only three species are common to both faunas.

Related soil types under different conditions of vegetation, weathering and podzolization can also be compared, and the effect of the soil processes upon the character of the fauna can be judged. With increased leaching and podzolization for example the total number of species in the soil tends to fall but this reduction appears to affect the ciliate fraction less than the rhizopod fraction. Again, wherever there is accumulation of litter or plant matter associated with acidity, rhizopods, especially testaceans, occur in large numbers.

There are some cases where not only the faunal pattern but also the species involved are identical for soils of similar type. Thus a ciliate, Phacodinium metschicoffi, which was recovered from a beech litter from Curious Cove, Marlborough Sounds, was again recorded from beech litter from the Taupo area. Again several testacean species found in sphagnum and peat of Campbell Island were encountered in sub-alpine soils from the Ruahines. In this case then not only did the faunal pattern of the sub-antarctic soils resemble that of the sub-alpine soils but species found nowhere else were common to both faunas.

Role of Protozoa in the Soil

Like all biological energy systems the soil organic cycle is an open system, that is, energy is constantly both being introduced and lost. That part of the organic cycle with which the protozoa are concerned takes place within the interstices and pore spaces between the various soil particles, organic and inorganic. Here in the microscopic architecture within and between soil particles or in holes formed by worms or plant roots and fissures left where the soil has parted live the soil micro-organisms. Bacteria, actinomyces, fungi and algae represent the plant kingdom; protozoa, mematodes, rotifers, gastrotrichs, tardigrades, enchytraeids (pot worms), mites and insect larvae represent the animal kingdom. To all these organisms moisture is essential for their activity and this is retained in the spaces from drainage loss partly by capillary action and partly by the absorbtive power of the soil colloids. As the soil dries out the colloids form a thin film around the particles binding them together.

It is for this reason that protozoa and other micro-organisms are not immediately visible if soil is examined directly under the microscope. When, however, a drop of water is added to the soil and it is examined some hours later active protozoa and other organisms will be found. Because the first ciliate protozoa to be observed appear to have recently excysted, which they are able to do within several hours, it has generally been assumed that protozoa are not active when confined to the colloidal film. When sufficient moisture is present however they may be seen active in the pore spaces and fissures where the water is held for a time.

Protozoa are capable of using a wide range of food sources. They may live on dissolved carbohydrates and proteins, on bacteria, on other protozoa whether of the same or different species or on organic debris formed by plants or fungal mycelia. Their powers of synthesis are however limited and generally like higher animals they require vitamins or growth factors for normal growth, besides a carbon and nitrogen source. Recently Faure-Fremiet has described a typical association of micro-organisms, in this case the sulphur bacterium, Beggotia, and its protozoan predators. The bacterium grows in a thick mat on bottom mud deriving its energy from the oxidation of hydrogen sulphide. When a colony develops it is invaded by a succession of protozoa, some feeding on the bacteria and some on each other, until finally the colony is consumed.

Protozoa in the soil are largely predators on bacteria, and it has been shown that there is an inverse ration between the numbers of the two groups of organisms. Because of this relationship it was once thought that protozoa were responsible for limiting bacterial activity. At the turn of the century the beneficial effects of partial sterilization of soil by steam or chemicals was realized. Russell and Hutchinson suggested that these beneficial effects were due to the removal of a factor limiting bacterial activity, and consequently plant growth, and that this factor was the protozoan fauna. Recently samples of steam sterilized soil were examined to investigate the effect on the protozoan fauna. The sterilized top-soil was found to have only 40% of the species present in the unsterilized top-soil but the number of active protozoa present in the sterilized soil was far greater than in the unsterilized.

From these results it is obvious that while Russell and Hutchinson were correct in surmising that the number of protozoan species would be drastically reduced, they were in error when they attributed the heightened activity of the bacteria to the absence of protozoa. For not only are protozoa present in the sterilized soil but they are present in very great numbers and their activity is stimulated by the sterilization just as much as the activity of the bacteria. Both show an initial inhibition of growth followed by greatly stimulated growth of that part of the fauna and flora which was not killed by steaming.

It is now recognized that the effect of steaming soil is more extensive than its biological effects and that physical and chemical characteristics of the soil are also affected. These include water-holding capacity, capillary properties, destruction of the colloidal film surrounding soil particles, increased concentration of the soil solution, and increased solubility of soil constituents, e.g. Ca, Mg, Mn, and phosphate and also to a lesser extent K, Si, and Al. The biological effects are an initial inhibition both of bacterial activity and seed germination followed by greatly stimulated bacterial and plant growth. The present view of sterilization is that the destruction of organisms removes possible disease-causing or antibiotic organisms and releases plant nutrients both by the autolysis of microbial protoplasm and by reducing microbial competition for nutrients. Deficiencies of trace elements can be removed by partial sterilization. The whole phenomenon may be viewed as a gross reorganization of the soil equilibrium and the stimulated bacterial, protozoan, and plant activity as resultants of the new equilibrium reached.

The view of Russell and Hutchinson then which treats bacterial activity as an inverse function of protozoan activity is erroneous. It is based upon a misconception of bacterial-protozoan relations. Both are directly affected by the soil equilibrium and any change in it which may occur. Although protozoa will undoubtedly reduce the number of bacteria in an environment they are unlikely to permanently affect the pulse of bacterial activity since any marked reduction of the bacterial flora will result in the death or encystment of the protozoa followed by renewed bacterial activity. Further by constantly preying on the bacterial flora they maintain it at the highest state of efficiency. They flocculate and condense dispersed particles and bring together into their own bulk energy sources which can then be used by larger organisms such as nematodes or else form centres of activity for fresh bacterial colonies.

Physiological Character of Soil Protozoa

Soil protozoa are characteristically small. Few species are over 100μ in size and over half of those most common in New Zealand soils are 50μ or less. This enables them to survive in the smallest drops of moisture and the thinnest moisture films. They are very hardy and can withstand extreme conditions of temperature, dessication, low oxygen and high carbon dioxide tensions. Their resistance to unfavourable conditions depends largely upon their ability to encyst. In the ciliate genus Colpoda, one of the most common and characteristic soil protozoa, there are basically two types of cyst. In one, under favourable conditions, cell division takes place, two, four or eight daughter cells being formed. The second type of cyst is a protective cyst formed in the absence of food and by crowding of the ciliates. Encystment may take place spontaneously, as with the reproductive cyst; or when the cysts are washed in a hypotonic solution such as distilled water or in response to some stimulant such as bacteria, hay infusion, yeast extract or alcohol. These protective cysts may be dried, heated to high temperatures (100° C.) or kept below freezing point and are still viable after months or even years of inactivity.

In the presence of food the soil protozoa can grow very rapidly. The reproductive rate of Colpoda steinii at room temperature is just under 4 per diem and of Vorticella microstoma, a wide-spread soil and sewage form, about 8. Not all bacterial strains are suitable as food for protozoa. So it is found that if loops of a suitable food bacterium are added to a culture of mixed bacteria and a single protozoan species (e.g. Colpoda steinii) the protozoan eats only the edible bacterium and consequently the non-edible bacteria increase in numbers. In this way protozoa have a selective effect on a mixed bacterial population. When only inedible bacteria are present the protozoa will encyst.

The flocculation of bacterial cells and other particles by protozoan activity has been recorded for a number of species, including the soil flalgellate Oikomonas termo and the coprophilic ciliate Balantiophorus minutus, also recorded from soil. In this latter case flocculation is achieved by mucus secreted by the ciliate.

In the complete absence of oxygen some free-living protozoa die within an hour or two, e.g. the fresh-water ciliate Stylonychia mytilus. Some survive for a time but do not feed, e.g. the normally sessile Vorticella microstoma which becomes free-swimming. Others feed normally under such conditions and can therefore survive indefinitely although the cells do not divide. To this last group belong Colpoda steinii and Trichopelma sphagnetorium, two common soil ciliates. Some protozoa are very sensitive to high carbon dioxide tensions whereas others such as Colpoda steinii and Colpoda inflata are very resistant. These two factors, oxygen lack and high carbon dioxide tension, are particularly relevant to soil conditions for they are both liable to occur in the capillary spaces where the protozoa live.

It can be seen from these few examples that the systematic distinction of the soil protozoan fauna is paralleled by morphological and physiological characteristics which make them particularly well suited for life in the soil. In all they combine a marked tolerance of varying environmental conditions with the ability to react quickly to favourable or unfavourable factors. These same characteristics favour them in other similar habitats and coprophilic and sewage protozoa consist largely of typical soil species. It is also interesting to find that the soil ciliate Colpoda steinii is an accidental parasite of slugs and snails, inhabiting particularly the pulmonary cavity where it divides without forming its typical reproductive cyst. Infection takes place through the slug eating material covering or containing the ciliate cysts and the parasite eventually causes the death of its host. This is a rare record of accidental parasitism by a free-living ciliate and is an indication of the adaptability of the protozoa.

Protozoa and Plant Growth

All nutrients that plants absorb have to pass a region of intense interactions between roots, microorganisms and animals, termed the rhizosphere. Plants allocate a great portion of their photosynthetically fixed carbon to root-infecting symbionts, such asmycorrhizal fungi; another part is released as exudates fuelling mainly free-living rhizobacteria. Rhizobacteria are strongly top-down regulated by microfaunal grazers, particularly protozoa. Consequently, beneficial effects of protozoa on plant growth have been assigned to nutrients released from consumed bacterial biomass, that is, the 'microbial loop'. In recent years however, the recognition of bacterial communication networks, the common exchange of microbial signals with roots and the fact that these signals are used to enhance the efflux of carbon from roots have revolutionized our view of rhizosphere processes. Most importantly, effects of rhizobacteria on root architecture seem to be driven in large by protozoan grazers. Protozoan effects on plant root systems stand in sharp contrast to effects of mycorrhizal fungi. Because the regulation of root architecture is a key determinant of nutrient- and water-use efficiency in plants, protozoa provide a model system that may considerably advance our understanding of the mechanisms underlying plant growth and community composition.

The Rhizosphere: Interface of Microbial and Faunal Interactions

A century ago go Hiltner introduced the term 'rhizosphere' to describe the stimulation of biomass and activity of microorganism in soil around plant roots. However, even today it is often not fully acknowledged that all nutrients a plant absorbs from soil pass through a region of intense microbial and faunal activity.

The stimulation of microbial activity in the rhizosphere results from the fact that plants secrete an array of low- and high-molecular weight molecules into the soil as exudates, which may account for up to 40% of the dry matter produced by plants. As free-living soil microorganisms are strongly carbon limited, a specialised microflora, typically consisting of fast growing bacteria, is triggered into activity by the carbon pulses provided as exudates. Root-derived carbon leads to strongly increased levels of microbial biomass and activity around roots and channels energy to subsequent microfaunal grazers, where numbers of bacterial feeding protozoa and free-living nematodes may increase up to 30-fold compared with bulk soil.

Estimates of plant-below-ground investments vary widely, but even if the C-transfer to exudation was 10–20% of total net fixed carbon, other microbial symbionts such as mycorrhizae or N_2-fixing microorganisms may each consume another 10–20% of total net fixed carbon. Although a trade-off between plant C-investment in different microbial interactions has been observed, plants may still release up to half of their total fixed carbon to fuel microbial interactions in the rhizosphere.

It becomes immediately clear that supporting microbial interactions in the rhizosphere must be of fundamental importance for plants to justify this significant trade-off in carbon allocation, which could otherwise be used to construct light-capturing or defensive structural tissues above-ground. In particular, why are plants providing ample energy in form of exudates to a microbial community that is strongly competing with roots for available nutrients? The answer partly lies in the loop structure of the bacterial energy channel in the rhizosphere. Nutrients become only temporarily locked up in bacterial biomass near the root surface and are successively liberated by microfaunal grazing. The interplay between microorganisms and microbivores determines the rates of nutrient cycling and strongly enhances the availability of mineral nutrients to plants. The assumed mechanism, known as the 'microbial loop in soil', is triggered by the release of root exudates from plants that increase bacterial growth in the rhizosphere. Plant available nutrients will be strongly sequestered during microbial growth and would remain locked up in bacterial biomass if consumption by protozoa and nematodes would not constantly re-mobilize essential nutrients for plant uptake. Due to the relatively small differences in respect to C : N ratios between predators and bacterial prey and a relatively low assimilation efficiency, only 10–40% and 50–70% of the prey carbon will be used for biomass production by protozoa and nematodes, respectively. The excess N is excreted as ammonia and hence is readily available for other soil organisms, including plant roots.

In addition, the populations of soil protozoa strongly fluctuate through time and parallel to the decline in protozoan numbers their rapidly decomposable tissue may enter the detrital food-web. Although the size of most protozoa in soil may range only between 10 and 100 μm in diameter, protozoan biomass is anything but small. In most soils protozoan biomass equals or exceeds that of all other soil animal groups taken together – with the exclusion of earthworms. Roughly estimated, 70 and 15% of total respiration of soil animals might be attributed to protozoa and nematodes, respectively. High production rates of protozoa with 10–12 times their standing crop per year and minimum generation times of 2–4 h suggest a strong grazing pressure on bacterial biomass and subsequent significant effects on nutrient mineralization.

Microfauna and Plant Growth

Beneficial effects of protozoa on plant growth are well documented. Experiments in planted microcosms provided strong evidence of the importance of protozoan grazing in the rhizosphere. Shoot biomass and amounts of shoot N mostly increased in the presence of protozoa and nematode grazers. Microfaunal stimulation of nitrogen mineralization via the microbial loop was suggested as the main underlying mechanism.

Consequently, food-web models simulating N-mineralization in soil suggest protozoa and bacterivorous nematodes to be the most important contributors to nitrogen mineralization. Their indirect contributions to nutrient cycling may be even more important than their direct effects because grazing stimulates microbial mineralization processes. For example, calculated that the contribution of amoebae and nematodes to overall N mineralization in winter wheat was 18 and 5%, but their subsequent deletion from the food-web model resulted in reductions of 28 and 12% of N mineralization for amoebae and nematodes, respectively.

Although nutrient-based models appear fully sufficient to estimate the gross outcome of plant–microfauna interactions, it must be noted that the underlying mechanisms are much more complex than could have been imagined nearly 20 yr ago when the microbial loop concept was first proposed.

Roots do not only secrete carbon; the rhizodeposition of nitrogen by plants can be substantial. By including the amount of nitrogen lost through rhizodeposition and modeling N transformations in the rhizosphere, and calculated that plant-derived carbon is only sufficient to allow for recycling of the N lost from the plant by exudation rather than to support mineralization of N from soil organic matter. Therefore only a small benefit of direct microfaunal activity to the gross N nutrition of plants can be assumed.

Indeed, protozoa have been found to increase plant biomass independently of nutrient contents in plant tissue. In a laboratory experiment, even a constant surplus of nutrients did not prevent an increase of up to 60% in biomass of spruce seedlings in the presence of protozoa, but completely eliminated beneficial effects of mycorrhiza.

These discrepancies in the microbial loop model suggest additional, nutrient-independent effects of protozoa on plant growth.

Victims and Benefactors: How Grazers affect Bacterial Community Composition

Bacteria in the rhizosphere are strongly top-down regulated by grazing and there is increasing evidence that changes in the composition of rhizobacteria due to strong and selective grazing are a major determinant of microfaunal effects on plant growth.

Protozoan effects on root architecture and composition of bacterial communities in the rhizosphere of rice (Oryza sativa). (a,b) Differences in root architecture of 16-d-old rice seedlings growing in Petri dishes on agar inoculated with a diverse soil bacterial community in absence (left) and presence (right) of amoebae (Acanthamoeba sp.). The length of the white bar is 1 cm. (b,c) An example for grazing-induced shifts in bacterial communities. Fluorecscent in situ hybridization of bacteria (red: alpha-proteobacteria, blue: eubacteria) on the agar surface near lateral roots of the plants above in absence (left) and presence (right) of amoebae. White arrows indicate red-fluorescent amoebal cysts. Please note the decrease of colonies of alpha-proteobacteria in presence of amoebae.

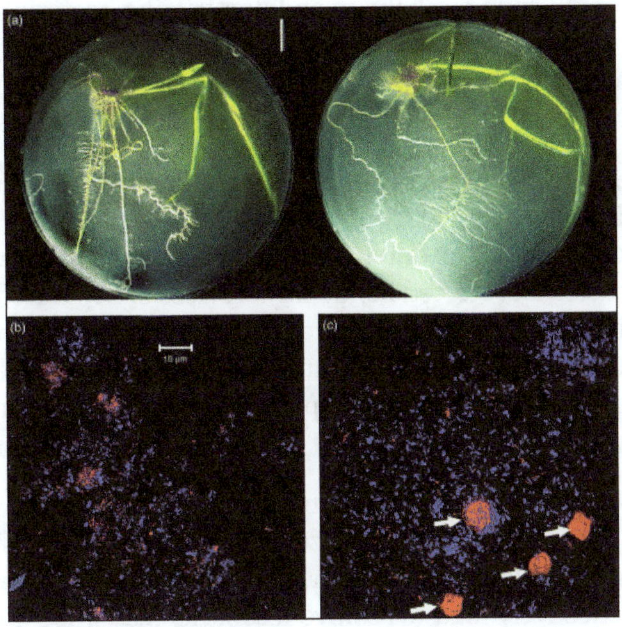

The most important bacterial grazers in soil are naked amoebae due to their high biomass and turnover and specialised feeding modes. By contrast to suspension and filter feeders like bacterivorous nematodes and other protozoa, amoebae are grazing bacterial biofilms and colonies attached to soil and root surfaces and thus have access to the majority of bacteria in soil. With the aid of their pseudopodia, amoebae can reach bacterial colonies in soil pores and even inside roots inaccessible to other predators and they may still continue grazing in tiny water films when other protozoa or nematodes are restricted by decreased water potential in soil. 'Bacterial' feeding nematodes, by contrast, are able to migrate to places of high bacterial and protozoan activity where they aggregate and become important grazers of both bacteria and amoebae and eventually become the dominant predators.

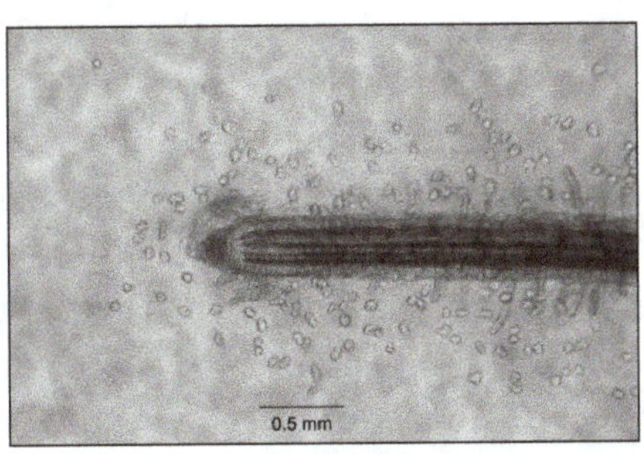

Typical distribution of amoebae (Acanthamoeba castellanii) along a lateral root of rice (Oryza sativa) growing on agar.

Significant changes in bacterial diversity due to protozoan grazing have been confirmed in freshwater systems as well as in the rhizosphere of plants. The grazing-induced changes in microbial functioning affect fundamental ecosystem properties because bacteria in soil occupy some of the most important control points for nutrient cycling and plant growth. For instance, N_2-fixing, nitrifying and denitrifying bacteria are in command of the nitrogen cycle. Protozoan grazing does often stimulate nitrifying bacteria, presumably through predation on their faster-growing bacterial competitors, resulting in high concentrations of NO_3^- in culture liquid and leachate of rhizosphere soil.

By contrast to interactions in freshwater systems, the three-dimensional structure of the soil habitat adds to the complexity of trophic interactions in the rhizosphere. Bacteria in biofilms on roots and on the outer zones of soil particles may receive greater grazing pressure than bacteria protected inside tiny crevices or separated by water films. In addition, temporal dynamics of bacterial activity on roots have to be considered. The heterogeneity of the rhizosphere in space and time needs considerably more attention in order to understand the contribution of predator–prey interactions to the dynamics of rhizosphere processes.

Bacterial communities often respond with morphological shifts to grazing which are directed either towards larger or smaller cell sizes. Protozoan grazing on bacteria commonly results in an outgrowth of filamentous bacterial cells and microcolonies that the protozoa apparently cannot effectively ingest. Positive feedbacks may arise through the removal of senescent bacteria and an increase in the contribution of younger strains with higher metabolic activity. Subsequent bottom-up effects occur through enhanced substrate availability, favoring those species capable of balancing their predation losses with enhanced growth rates. However, the bigger cells of these actively growing bacterial populations seem most attractive to grazers and again receive the largest grazing pressure. For example, grazing of protozoa in the rhizosphere of wood barley (Hordelymus europaeus) resulted in a 20–40% increased respiratory quotient, qO_2 (i.e. microbial respiration per unit microbial biomass), confirming a strongly enhanced turnover of the grazed bacterial community and evolution of $^{13}CO_2$ from microbial mineralization of labeled plant litter material increased up to 100% in the rhizosphere of ryegrass due to protozoan grazing.

However, not only bacterial size and turnover matters. The fact that certain microfaunal grazers grow better on some bacterial species than on others is well established for protozoa and nematodes. There is strong evidence that nematodes and protozoa use chemical clues to sense and discriminate between their bacterial prey species. In the case of protozoa a high selectivity in bacterial food choice has been confirmed and not only the feeding strategies and grazing rates may considerably differ between taxa and species, but also the sensitivity of the bacterial prey to grazing may strongly diverge and depend on the protozoan predator. For example, the same bacterial prey species may be digested differently by various protozoan grazers, and the same predator species may selectively digest variable prey. In addition, some common pigmented soil bacteria, like Chromobacter, evolved secondary compounds with extreme toxicity to protozoa.

Studies in experimental soil systems indicate that gram-positive bacteria benefit while gram-negative bacteria decrease by grazing. However, demonstrated that not all gram-negative bacteria were reduced, for example Pseudomonas, a typical rhizosphere colonizer, increased during grazing. Moreover, their study demonstrated that the grazing-induced changes in bacterial diversity were partly dependent on the taxa and species of protozoan grazers present. Thus, grazing of protozoa, and possibly nematodes, creates a complex top-down pressure affecting the morphological, taxonomic and functional composition of the bacterial community.

Molecular techniques, such as fluorescent in situ hybridization, microautoradiography, stable isotope probing and metagenomics now offer ways to directly visualize changes in the identity, activity, function and spatial arrangements of bacterial communities exposed to protozoan predation. They may considerably advance our understanding of the role of microfaunal predators in affecting the function and spatial organization of bacterial communities in the rhizosphere of plants.

Nonmotile Plants with Flexible Strategies

Plants are not passive recipients of nutrients, instead plants integrate information from the environment into their decisions on below-ground investments such as root production and proliferation, formation of symbiotic relationships with infecting microorganisms (e.g. mycorrhizal fungi, or N_2-fixing bacteria), alteration in exudation rates, interactions with free-living bacteria or production of secondary compounds to defend herbivores. Because root morphology is both genetically programmed and environmentally determined, there must be signal transduction pathways that interpret complex environmental conditions and activate genes to enter a particular symbiosis or to form a lateral root at a particular time and place. Microbial symbionts must communicate their presence to plant hosts and plants need to distinguish friend (symbiont) and foe (pathogen), suggesting that the reciprocal exchange of microbial signals with roots is common. Recently, combined these insights in the concept of 'rhizosphere' control points' in order to emphasize the importance of information exchange between plants and microorganisms for gene expression patterns and resulting morphological and physiological changes in the partners.

From a microbial perspective the evolution of strategies enhancing energy transfer to the roots led to an increase in fitness of those microorganisms that influence gene regulation in plants by sending the respective signals. Indeed most of the specialised rhizosphere bacteria appear to have the potential to affect plant performance by producing hormones. Up to 80% of the bacteria isolated from plant rhizospheres are considered to produce auxins, and their widespread ability to produce cytokinins led to suggest that cytokinins in plants may originate exclusively from microorganisms. The widespread ability of both beneficial and deleterious rhizosphere organisms to produce plant hormones and other signal molecules suggest that rhizosphere bacteria play an important role in manipulating root and plant growth.

Recent advances in the application of microbial biosensors are expected to significantly advance our understanding of the spatial context of substrate availability and signal exchange at scales relevant to roots and microorganisms.

References

- Actinomycetes: the-compost-gardener.com, Retrieved 13 July, 2020
- An-introduction-to-actinobacteria, actinobacteria-basics-and-biotechnological-applications: intechopen.com, Retrieved 23 April, 2020
- Rhizobium: microscopemaster.com, Retrieved 13 May, 2020
- Harmful-bacteria-in-soil-13406918: hunker.com, Retrieved 27 March, 2020
- Pathogenic-bacteria-soil-8346272: ehow.com, Retrieved 24 February, 2020
- Soil-Algae, soil-7-microorganisms-found-in-soil-microbiology-66603: biologydiscussion.com, Retrieved 16 July, 2020
- Agricultural-importance-of-algae-282791348: researchgate.net, Retrieved 23 May, 2020

Soil Environments 4

- **Soil Habitat and Soil Ecology**
- **Factors Affecting Microbial Community in Soil**
- **Relationship of Agronomic Practices to Soil Nitrogen Dynamics**

Soil environment means the physical, biological and chemical factors that affect the microbes present in the soil. Soil habitat, soil ecology, microbial community, soil nitrogen dynamics, etc. are studied under its domain. The topics elaborated in this chapter will help in gaining a better perspective of soil environment.

The soil environment consists of a variety of physical, biological and chemical factors that affect the abundance and diversity of microbes found in the soil. At its basic level, the soil environment consists of a solid and porous fraction. Within these fractions, a variety of chemical and physical factors are affected by and and affect microbes. These include, but are not limited to texture, temperature, pH, oxygen, cation exchange capacity and redox reactions.

The soil environment directly affects the types of microbes, as well as the rates of processes they perform. For example, microbial activity increases with temperature, which in turn affects rates of decomposition. On the other hand, microbial processes directly affect their environments as well, contributing to the carbon and nitrogen cycles, which are important for microbial and plant health. At the micro scale, bacteria and other microbes participate in a variety of reactions that affect nutrient cycling, pH, as well as oxygen and CO_2 content. At the macro scale, these processes can change the landscape in drastic ways, assisting in weathering of the soil and development of soil layers.

Solid Fraction

The solid fraction of the soil consists of mineral and organic matter, which is typically about 50% of the soil by volume, and it has a dominant influence on heat, water, and chemical transport and retention process. Most of the solid particles are derived from mineral sources such as decomposed rocks or sediments. Soil organic matter (SOM) consists of all of the organic components of a soil, including living biomass, decomposing tissue, and fully decomposed tissue called humus. The rate and pathway of carbon decomposition and SOM formation directly affects the carbon cycle. Texture can also influence chemical properties such as cation exchange capacity (CEC). Finer textured soils with high clay content will have great CEC than soils with low clay content.

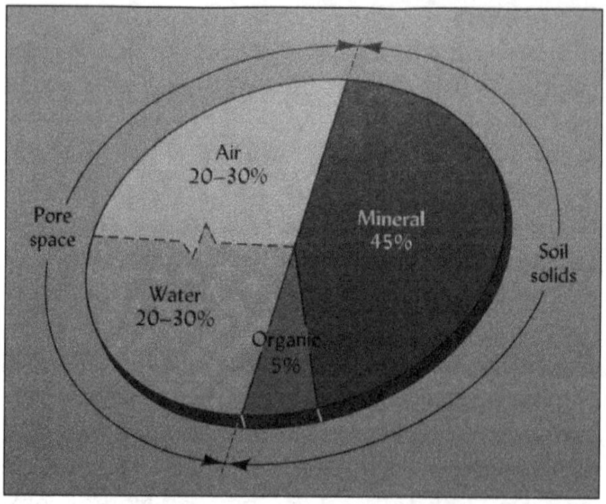

Volume composition of the solid and porous fractions of a typical loam surface soil.

Soil Pores

Soil pores consist of the air and water filled fractions of the soil, and together they make up about 50% of the soil by volume. Pore space is largely determined by size and arrangement of aggregates and affects the movement of water, air, and organisms in soil. Soil pores are typically classified based on size: Macropores (>75µm) Mesopores (30-70µm) Micropores (5-30µm) Ultramicropores (0.1-5µm) Cryptoporus (<0.1µm).

The air filled pores of the soil typically have a similar distribution of gases as the atmosphere above the soil, with slightly lower oxygen and slightly more CO_2 due to the respiration of microorganisms. The soil atmosphere consists of about 18-20% oxygen near the surface, which decreases with depth. CO_2 is around 1%, and N_2 is about 78% of the soil air filled pore space. Oxygen content will be lower when available carbon is high (demand for high O_2 to utilize carbon). Soil that is high in clay content and compacted may have trouble exchanging gases to the atmosphere. Soil air generally has a very high moisture content when compared to the atmosphere (~100% unless the soil is very dry). The amount and composition of air in a soil is largely determined by the water content in the soil.

The main source of soil water is rainfall and overland flow. The amount of water that enters the soil is a function of soil structure and texture. Water moves in soil through mass flow and capillary action. The water in soil is often called the soil solution, which can move nutrients from the surface through the soil column. The water fraction of the pores is typically between 20% to 30% but can vary depending on precipitation, soil texture, and soil structure. The water content of a soil influences gas exchange, nutrient movement and concentrations of nutrients, and buffers the temperature of the soil.

Soil Aggregates and Structure

Soil aggregates are groups of soil particles that bind to each other more strongly than to adjacent particles. The space between the aggregates provide pore space for retention and exchange of air and water. Aggregation affects erosion, movement of water, and plant root growth. Polysaccharides produced by soil bacteria, and humic substances and hyphae produced by fungi improve aggregation.

A soil aggregate at the micro and macro scale.

Soil structure is the arrangement of primary soil particles into aggregates which describes the arrangement of the solid parts of the soil and the pore spaces between them. Soil structure has a major influence on water movement, SOM leaching, and gas exchanges of the soils. The water movement, affected by structure can bring SOM in the surface to deep inside the soil. Soil with high clay content and poor structure, may have reduced infiltration and will cause runoff, erosion and surface crusting. This can cause nutrient loss and increase the potential of desertification.

Physical and Chemical Factors that Control Biological Activity in the Soil

Texture

Soil texture is defined as the distribution of sand (0.05-2.0 mm), silt (0.002-0.05mm), and clay (< 0.002mm) in soil. Soil texture indirectly influences properties such as: water holding capacity, porosity, aeration and nutrient availability. Clay particles have a very high surface to volume ratio, which makes them very chemically active and have high nutrient availability. Due to the adhesion of water, soils high in clay will also have a high water holding capacity. Soils with high clay content will often have a very active microbial community, especially in areas of the rhizosphere.

One study by Hamarshid found that CO_2 production rates were greatest in fine textured soil compared to coarse textured soil. This was due to finer textured soils' greater ability to hold nutrients and water, allowing microbial populations to thrive. However, this is not always the case, as other chemical or physical properties may affect the ability of microbes to perform processes such as carbon decomposition. (e.g. temperature, pH and quality of substrate).

Temperature

Soil temperature changes with depth: The surface soil (~0-20 cm) is highly affected by the solar radiation. Moving deeper (~below 27 cm) temperatures are very stable over time. This is because heat moves in soil mainly by conduction, which does not allow much heat to reach deep in the soil profile. Soil temperature is also affected by the soil color, soil cover, and the water content of the soil. A darker soil can absorb more heat compared to lighter color soil. A dry soil is more easily heated than a wet one due to the higher heat capacity of water. Heat moving in soil is analogous to the movement of water. Generally, the higher the temperature, the more active microbes are, with microbial activity typically doubling with a 10° rise in temperature. However, some bacteria thrive at very low temperatures (psychrophiles) and very high temperatures (extremophiles).

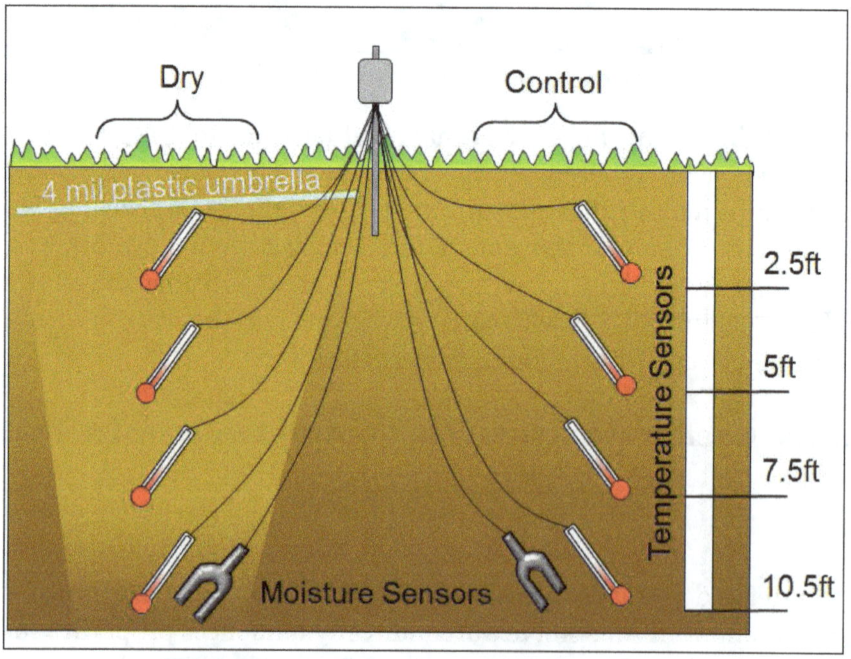

Soil temperature fluctuation as a function of depth.

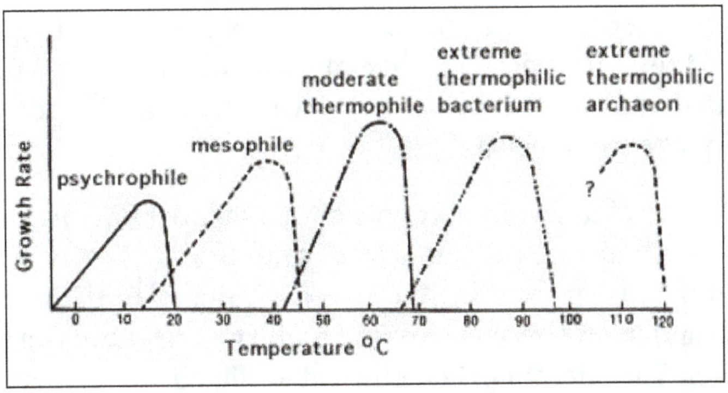

Microbial growth rate as a function of temperature for different types of bacteria.

pH

pH change in soils is due to both biotic and abiotic processes. Microbes consume and release H^+ through redox reactions and fermentation. Abiotic processes such as rainfall can also affect the pH of the soil. In areas of high rainfall, acidic soils can be created through leaching of bases from the soil, while more basic soils are typically located in arid environments. One study by Fierer et al, found that pH was the single most important factor influencing microbial diversity at the continental scale. The study used 98 samples from across North and South America, and found that the greatest microbial diversity coincided with a neutral pH, even between two soil from the same forest.

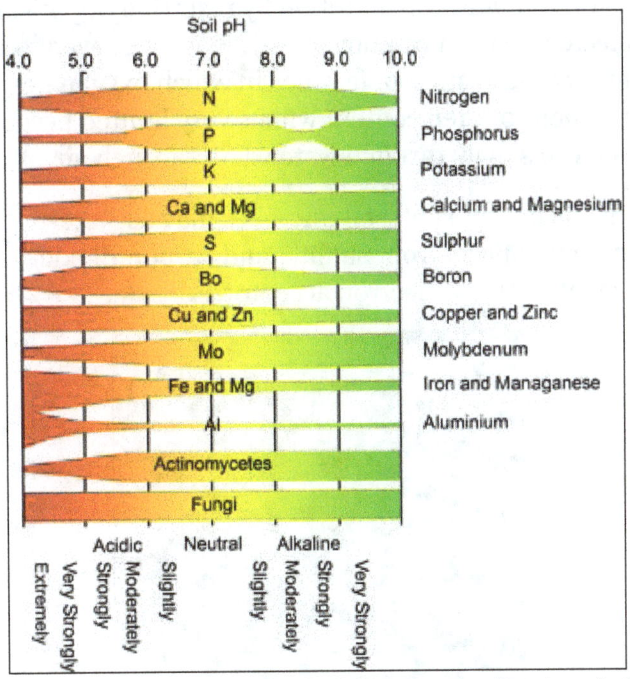

Nutrient availability and microbial diversity as a function of pH.

pH affects microbial diversity because many microbial species cannot tolerate extreme levels of pH (high or low). Alterations in pH can render essential microbe enzymes inactive and denature proteins within the cells and prevent microbial activity from occurring. However, there are microbes that can withstand extreme pH environments. At pH below 5, fungi and acidophilic bacteria have a competitive advantage over other bacteria that thrive at a more neutral pH.

pH can also affect the availability of nutrients in this soil. Below a pH of 5, essential plant nutrients such as phosphorous, calcium and magnesium are not available. Low pH can also cause aluminum (Al_3^+) to be released from soil minerals. Al_3^+ in soil solution is not only toxic to plants and microbes, it can combine with OH^- ions causing the free H^+ ions to lower the pH further. The effect of pH on nutrient availability and microbe survivability.

Oxygen

Oxygen (O_2) is a very important component of the productivity of both microbes and plant roots. Oxygen has a very high electrical potential (Eh), meaning that it has a lot of potential to produce energy when used as an electron acceptor in an oxidation-reduction reaction. An example oxidation-reduction reaction can be seen in equation 1, where glucose is being oxidized, and oxygen is being reduced.

$$6O_2 + C_6H_{12}O_6 \rightarrow 6CO_2 + 6H_2O$$

The amount of oxygen available in a soil depends on a number of factors, including soil porosity, water content, and consumption by respiring organisms. If soil pores are large and interconnected, oxygen can flow easily. However, even in a well aerated soil, micro-aggregates may contain anaerobic zones in which oxygen flow is very limited. In a flooded soil environment oxygen content will be very limited because oxygen diffuses about 10,000 times more slowly through water than through air.

In areas where oxygen is not present, soil microbes may use alternative electron acceptors such as nitrate, manganese, iron, sulfate, and carbon dioxide. An example of the result of the use of an alternative electron acceptor.

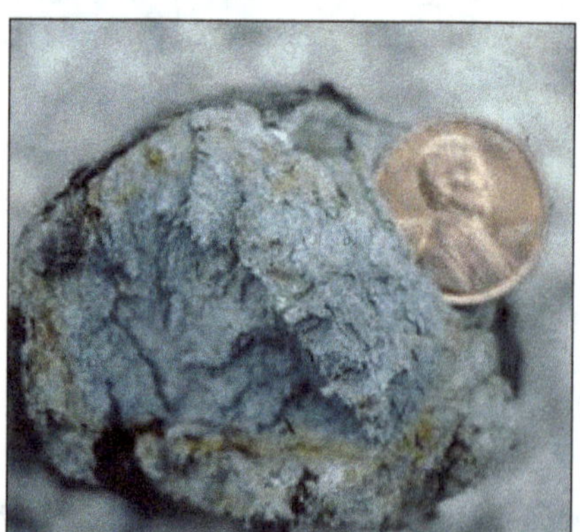

Without oxygen, bacteria can potentially use iron as an electron acceptor, reducing it, and creating the blue/gray soil in the photo. The mottled red areas are the result of oxygen getting in (perhaps through root canals) and oxidizing the iron.

Cation Exchange Capacity

Cation exchange capacity (CEC) is the ability of a soil to hold and exchange cations. The amount of CEC in soil is highly dependent on the texture and organic matter of the soil. The high surface area and negative charge of clay allows it to bind and exchange with soil solution, which contains cations that are important for plant and microbial health.

For many soils, especially those with low clay and organic carbon content, CEC is dependant upon soil pH. As pH decreases, an increasing amount of H^+ ions are attracted to negatively charged clay particles and functional groups in SOM. This causes other cations, that were attached to these surfaces to fall off into the soil solution. CEC will increase as pH increases as less cations are being pushed out by H^+ ions.

At typical soil pH values (5-8), microbes and clay particles are both negatively charged, but microbes still to bind to clay. This binding could be due van der waals, hydrogen bonding, sharing electrons and ion exchange.

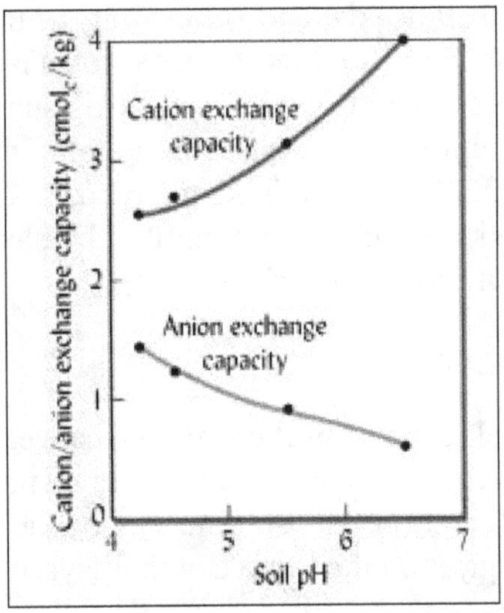

Cation exchange capacity and anion exchange capacity of a soil as function of soil pH.

Redox Reactions

Reduction-oxidation (redox) reactions are chemical reactions in which reactants experience a change in oxidation number (which means these reactants either gain or lose electrons). Many reactions in the soil involve the gain or loss of electrons, obtaining or releasing energy. Redox reactions are important in the soil, because microbes obtain energy through redox reactions for their metabolism, and the redox state can also determine the microbial processes that will occur.

Redox reactions include anabolism and catabolism process, both of which play important roles in microbial metabolism. Anabolism is the biosynthesis of cellular components, linked to energy requirements, while catabolism is the the biochemical processes leading to breakdown of organic substances, linked to energy production. For example, the photosynthesis and respiration processes are the coupled reactions in the soil, where plants required energy from light and reduce carbon dioxide to glucose, which then be used by microbes in the soil as the energy for their metabolism.

The tendency of compounds to accept or donate electrons is expressed as "redox potential". Redox potential (Eh), determined from the concentration of oxidants and reductants in the environment, is to measure the oxidation-reduction status and electron availability within this system. Electrons are essential to all inorganic and organic chemical reactions. Redox potential measurements allow for rapid characterization of the degree of reduction and for predicting stability of various compounds that regulate nutrients and metal availability in soil and sediment. The inorganic oxidants include oxygen, nitrate, nitrite, manganese, sulfate, and CO_2, while the reductants include various organic substrates and reduced inorganic compounds (NH_4^+, Fe_2^+, Mn_2^+, S_2^-, CH_4, and H_2).

The redox potential of a substance depends on Affinity of molecules for electrons and Concentration of reductants and oxidants (referred to as redox pair) Anaerobic environments such as wetland soils and flooded soils are usually limited by electron acceptors and have an abundant supply of electron donors. In this case, most microbes' activity is limited, and facultative and obligate microbes reduce the minerals following the electron tower. Aerated soils are usually limited by electron donors and have an abundant supply of electron acceptors (primarily O_2).

Salinity

Soil salinity refers to the salt content in the soil. The concentrations and types of ions in solution in the soil can cause modifications in the dispersion of the clay fraction, degrading the original soil fraction. The sodium ion, being monovalent, increases the width of the diffuse double layer on the surface of the clays, reducing the attractive forces between them with a consequent increase in particle dispersion. The consequence of this dispersion of the clay is also shown by a reduction in stability of the soil aggregates, which are thus easily transported by rain or irrigation.

Soil salinization is a big problem for soils in arid or semi-arid regions and agricultural soils throughout the world. Salts can adversely affect plant and microbial growth, due to destruction of the soil structure and its consequent compacting. The stress of high salt concentration can be detrimental for sensitive microorganisms and decrease the activity of surviving cells, due to the metabolic load imposed by the need for stress tolerance mechanisms.

Bioavailability

Bioavailability assesses what proportion of a contaminant present at a contaminated site is available for uptake by organisms. Bioavailability processes are the biological, chemical and physical processes that result in an organism being exposed to a contaminant present in the soil. These processes are: release of the contaminant from the solid phase, transport of the contaminant to and across a biological membrane and, incorporation into a living organism.

Bioavailable molecules must cross a biological membrane, which means the molecules have to interact with the aqueous phase. Therefore, soil properties which control partitioning between the solid phase in soil and the pore water, such as pH, organic matter content, Eh, cation exchange capacity (CEC), and the concentration of clay minerals, have a significant impact on bioavailability. Increasing exchange sites aids the retention of molecules in the pore water in a bioavailable form, but molecules sorbed strongly to surfaces or in solid form are generally not bioavailable.

Physical Properties of Soil

Horizonation

- Soil "horizons" are discrete layers that make up a soil profile. They are typically parallel with the ground surface. In some soils, they show evidence of the actions of the soil forming processes.
- Horizons are dominated by organic material. Some are saturated with water for long periods or were once saturated but are now artificially drained; others have never been saturated.
- A horizons are mineral layers that formed at the surface or below an O horizon, that exhibit obliteration of all or much of the original rock structure, and that show one or both of the following:
    - An accumulation of humified organic matter intimately mixed with the mineral fraction and not dominated by properties characteristic of E or B horizons.
    - Modification as a result of the actions of cultivation, pasturing, or similar kinds of disturbance.
- E horizons are mineral layers that exhibit the loss of silicate clay, iron, aluminum, humus, or some combination of these, leaving a concentration of sand and silt particles. These horizons exhibit obliteration of all or much of the original rock structure.
- B horizons are mineral layers that typically form below an A, E, or O horizon and are dominated by obliteration of all or much of the original rock structure and show one or more of the following:
    - Alluvial concentration of silicate clay, iron, aluminum, humus, carbonate, gypsum, or silica, alone or in combination.
    - Evidence of removal of carbonates.
    - Residual concentration of sesquioxides.

   - Coatings of sesquioxides that make the horizon conspicuously lower in value, higher in chroma, or redder in hue than overlying horizons without apparent illuviation of iron.

   - Alteration that forms silicate clay or liberates oxides or both and that forms granular, blocky, or prismatic structure if volume changes accompany changes in moisture content; or brittleness.

- C horizons are mineral layers which are not bedrock and are little affected by pedogenic processes and lack properties of O, A, E or B horizons. The material of C layers may be either like or unlike that from which the overlying soil horizons presumably formed. The C horizon may have been modified even if there is no evidence of pedogenesis.

- R horizons are layers of hard bedrock.

Transitional horizons are dominated by properties of one master horizon, but have subordinate properties of another. A/B and B/C are examples of transitional horizon designations.

Soil Color

- In well aerated soils, oxidized or ferric (Fe^{+3}) iron compounds are responsible for the brown, yellow, and red colors you see in the soil.

- When iron is reduced to the ferrous (Fe^{+2}) form, it becomes mobile, and can be removed from certain areas of the soil. When the iron is removed, a gray color remains, or the reduced iron color persists in shades of green or blue.

- Upon aeration, reduced iron can be reoxidized and redeposited, sometimes in the same horizon, resulting in a variegated or mottled color pattern. These soil color patterns resulting from saturation, called "redoximorphic features", can indicate the duration of the anaerobic state, ranging from brown with a few mottles, to complete gray or "gleization" of the soil.

- Soils that are dominantly gray with brown or yellow mottles immediately below the surface horizon are usually hydric.

- Soil color is typically described using some form of color reference chart, such as the Munsell Color Chart. Using the Munsell system, color is described in reference to the color's "hue", "value", and "chroma". Hue describes where in the color spectrum the soil color exists, which for soils includes the colors yellow, red, blue, green, and gray. Value describes the lightness of the color. Chroma indicates the strength of the color. In a Munsell notation, the color is written in the order hue-value-chroma. The color "5YR 4/3" is an example of a Munsell notation, where 5YR is the hue, 4 is the value, and 3 is the chroma.

Soil Texture

- Soil texture refers to the proportion of the soil "separates" that make up the mineral component of soil. These separates are called sand, silt, and clay. These soil separates have the following size ranges:
    - Sand = <2 to 0.05 mm.
    - Silt = 0.05 to 0.002 mm.
    - Clay = <0.002 mm.
- Sand and silt are the "inactive" part of the soil matrix, because they do not contribute to a soil's ability to retain soil water or nutrients. These separates are commonly comprised of quartz or some other inactive mineral.
- Because of its small size and sheet-like structure, clay has a large amount of surface area per unit mass, and its surface charge attracts ions and water. Because of this, clay is the "active" portion of the soil matrix.
- For all mineral soils, the proportion of sand, silt, and clay always adds up to 100 percent. These percentages are grouped into soil texture "classes", which have been organized into a "textural triangle".

Soil Textural Triangle

- Soil texture can affect the amount of pore space within a soil. Sand-sized soil particles fit together in a way that creates large pores; however, overall there is a relatively small amount of total pore space. Clay-sized soil particles fit together in a way that creates small pores; however, overall there are more pores present. Therefore, a soil made of clay-sized particles will have more total pore space than a will a soil made of sand-sized particles. Consequently, clayey soils will generally have lower bulk densities than sandy soils.
- Collectively, the soil separates of sand, silt, and clay are called the "fine-earth fraction", and represent inorganic soil particles less than 2mm in diameter. Inorganic soil particles 2mm and larger are called "rock fragments".
- When the organic matter content of a soil exceeds 20 to 35% (on a dry weight basis) it is considered organic soil material, and the soil is called an organic soil. As this material is mostly devoid of mineral soil material, they cannot be described in terms of soil texture. However, the following "in lieu of" texture terms can be used to describe organic soils:
    - "Peat": Organic material in which the plant parts are still recognizable.
    - "Muck": Highly decomposed organic material in which no plant parts are recognizable.

- "Mucky peat": Decomposition is intermediate between muck and peat.

Soil Structure

The soil separates can become aggregated together into discrete structural units called "peds". These peds are organized into a repeating pattern that is referred to as soil structure. Between the peds are cracks called "pores" through which soil air and water are conducted. Soil structure is most commonly described in terms of the shape of the individual peds that occur within a soil horizon.

Types of Soil Structure Graphic Example.

Graphic Example	Description of Structure Shape
	Granular – roughly spherical, like grape nuts. Usually 1-10 mm in diameter. Most common in A horizons, where plant roots, microorganisms, and sticky products of organic matter decomposition bind soil grains into granular aggregates.
	Platy – flat peds that lie horizontally in the soil. Platy structure can be found in A, B and C horizons. It commonly occurs in an A horizon as the result of compaction.
	Blocky – roughly cube-shaped, with more or less flat surfaces. If edges and corners remain sharp, we call it angular blocky. If they are rounded, we call it subangular blocky. Sizes commonly range from 5-50 mm across. Blocky structures are typical of B horizons, especially those with a high clay content. They form by repeated expansion and contraction of clay minerals.
	Prismatic – larger, vertically elongated blocks, often with five sides. Sizes are commonly 10-100mm across. Prismatic structures commonly occur in fragipans.
	Columnar – the units are similar to prisms and are bounded by flat or slightly rounded vertical faces. The tops of columns, in contrast to those of prisms, are very distinct and normally rounded.

Description of Structure Shape

Graphic Example	Description of Structure Shape
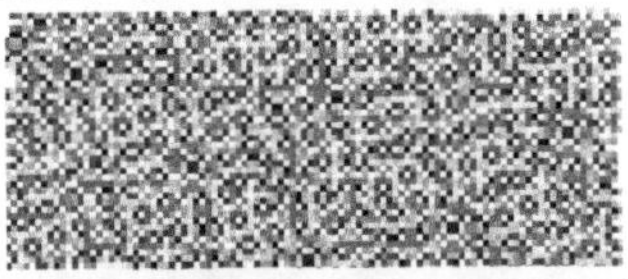	Massive – compact, coherent soil not separated into peds of any kind. Massive structures in clayey soils usually have very small pores, slow permeability, and poor aeration.
	Single grain – in some very sandy soils, every grain acts independently, and there is no binding agent to hold the grains together into peds. Permeability is rapid, but fertility and water holding capacity are low.

Soil Consistence

Soil consistence refers to the ease with which an individual ped can be crushed by the fingers. Soil consistence, and its description, depends on soil moisture content. Terms commonly used to describe consistence are:

- Moist soil:
    - Loose: Noncoherent when dry or moist; does not hold together in a mass.
    - Friable: When moist, crushed easily under gentle pressure between thumb and forefinger and can be pressed together into a lump.
    - Firm: When moist crushed under moderate pressure between thumb and forefinger, but resistance is distinctly noticeable.
- Wet soil:
    - Plastic: When wet, readily deformed by moderate pressure but can be pressed into a lump; will form a "wire" when rolled between thumb and forefinger.
    - Sticky: When wet, adheres to other material and tends to stretch somewhat and pull apart rather than to pull free from other material.
- Dry Soil:
    - Soft: When dry, breaks into powder or individual grains under very slight pressure.
    - Hard: When dry, moderately resistant to pressure; can be broken with difficulty between thumb and forefinger.

Bulk Density

Bulk density is the proportion of the weight of a soil relative to its volume. It is expressed as a unit of weight per volume, and is commonly measured in units of grams per cubic centimeters (g/cc).

Bulk density is an indicator of the amount of pore space available within individual soil horizons, as it is inversely proportional to pore space:

- Pore space = 1 – bulk density/particle density.

For example, at a bulk density of 1.60 g/cc, pore space equals 0.40 or 40%. At a bulk density of 1.06 g/cc, pore space equals 0.60 or 60%.

The addition of even a small percentage of organic soil material to a mineral soil can affect the bulk density of that soil. Compare the two soil samples below:

- Soil "A": 100% mineral soil material; bulk density = 1.33 g/cc.
- Soil "B": 95% mineral soil material and 5% organic soil material; bulk density = 1.26 g/cc.

The difference in bulk density relates to a difference in "particle density" of mineral soil material versus organic soil material. The average particle density of mineral soil material is 2.65 g/cc, which approximates the density of quartz. Conversely, the average particle density of organic soil material is 1.25 g/cc. Organic soil material weighs less than mineral soil material, so it will lower the bulk density of a mineral soil when added, as it reduces the overall weight of the soil.

Soil Chemical Properties

Cation Exchange Capacity (CEC)

Some plant nutrients and metals exist as positively charged ions, or "cations", in the soil environment. Among the more common cations found in soils are hydrogen (H^+), aluminum (Al^{+3}), calcium (Ca^{+2}), magnesium (Mg^{+2}), and potassium (K^+). Most heavy metals also exist as cations in the soil environment. Clay and organic matter particles are predominantly negatively charged (anions), and have the ability to hold cations from being "leached" or washed away. The adsorbed cations are subject to replacement by other cations in a rapid, reversible process called "cation exchange".

Cations leaving the exchange sites enter the soil solution, where they can be taken up by plants, react with other soil constituents, or be carried away with drainage water.

The "cation exchange capacity", or "CEC", of a soil is a measurement of the magnitude of the negative charge per unit weight of soil, or the amount of cations a particular sample of soil can hold in an exchangeable form. The greater the clay and organic matter content, the greater the CEC should be, although different types of clay minerals and organic matter can vary in CEC.

Cation exchange is an important mechanism in soils for retaining and supplying plant nutrients, and for adsorbing contaminants. It plays an important role in wastewater treatment in soils. Sandy soils with a low CEC are generally unsuited for septic systems since they have little adsorptive ability and there is potential for groundwater.

Soil Reaction (pH)

By definition, "pH" is a measure of the active hydrogen ion (H^+) concentration. It is an indication of the acidity or alkalinity of a soil, and also known as "soil reaction".

The pH scale ranges from 0 to 14, with values below 7.0 acidic, and values above 7.0 alkaline. A pH value of 7 is considered neutral, where H^+ and OH^- are equal, both at a concentration of 10-7 moles/liter. A pH of 4.0 is ten times more acidic than a pH of 5.0.

The most important effect of pH in the soil is on ion solubility, which in turn affects microbial and plant growth. A pH range of 6.0 to 6.8 is ideal for most crops because it coincides with optimum solubility of the most important plant nutrients. Some minor elements (e.g., iron) and most heavy metals are more soluble at lower pH. This makes pH management important in controlling movement of heavy metals (and potential groundwater contamination) in soil.

In acid soils, hydrogen and aluminum are the dominant exchangeable cations. The latter is soluble under acid conditions, and its reactivity with water (hydrolysis) produces hydrogen ions. Calcium and magnesium are basic cations; as their amounts increase, the relative amount of acidic cations will decrease.

Factors that affect soil pH include parent material, vegetation, and climate. Some rocks and sediments produce soils that are more acidic than others: quartz-rich sandstone is acidic; limestone is alkaline. Some types of vegetation, particularly conifers, produce organic acids, which can contribute to lower soil pH values. In humid areas such as the eastern US, soils tend to become more acidic over time because rainfall washes away basic cations and replaces them with hydrogen. Addition of certain fertilizers to soil can also produce hydrogen ions. Liming the soil adds calcium, which replaces exchangeable and solution H^+ and raises soil pH.

Lime requirement, or the amount of liming material needed to raise the soil pH to a certain level, increases with CEC. To decrease the soil pH, sulfur can be added, which produces sulfuric acid.

Soil Habitat and Soil Ecology

Soil is one of the more complex and highly variable habitats on earth. Any organisms that make their home in soil have had to devise multiple mechanisms to cope with variability in moisture, temperature, and chemical changes so as to survive, function, and replicate. Within a distance of,1 mm, conditions can vary from acid to base, from wet to dry, from aerobic to anaerobic, from reduced to oxidized, and from nutrient-rich to nutrient-poor. Along with spatial variability there is variability over time, so organisms living in soil must be able to adapt rapidly to different and changing conditions. Variations in the physical and chemical properties of the soil are thus important determinants of the presence and persistence of soil biota.

Differences among Soil Horizons

A typical soil profile has both horizontal and vertical structure. At the base of any soil profile is underlying bedrock, or parent material, which is the type of geological formation upon which and with which the soil above has been formed. Overlying the bedrocks a C horizon that has developed directly from modifications of the underlying parent material. This C horizon remains the least weathered (changed) of the identifiable horizons, accumulating calcium (Ca) and magnesium (Mg) carbonates released from horizons above. Microbial activity in this C horizon is typically very low, in part because of limitations in oxygen (O_2) and organic matter.

Overlying the C horizon is the subsoil, or B horizon. This is composed of minerals derived from the parent material and of materials that have leached down from the horizons above, including humid materials formed above from the decomposition of organic (plant and animal) matter. Yet, because the B horizon is typically still rather low inorganic matter, it supports relatively small microbial populations and has little biological activity. The B horizon is the zone of maximum illuviation, i.e., deposition or accumulation of silicate clays and of iron (Fe) and aluminum (Al) oxides.

The A horizon, denoting the upper layers of soil, is usually fairly high in organic matter and often darker in color. This, along with the O (organic) horizon, is the horizon in which plant roots and soil organisms are most active. Within the A horizon there are differing extents of leaching and movement of materials from the horizon above to the horizons below. The interface between the A and B horizons is the zone of maximum eluviation, i.e., removal through downward leaching of silicate clays and Fe and Al oxides. The interface between the A horizon and the O horizon above it is where incoming organic residues become incorporated with the mineral soil. Together with incorporated soil organic matter (SOM), the A horizon is often referred to as the topsoil.

The O horizon on the surface is the topmost layer, often referred to as the litter layer. The largest component of this layer is undecomposed organic matter (OM), and the origins of these organic materials are easy to distinguish — plant litter, manure, or other organic inputs.

Factors in Soil Genesis

In 1941, Hans Jenny proposed the following soil-forming factors that are still used today:

- The parent material or underlying geological formation of the region.

- The climate, referring largely to the temperature and precipitation in the region and to their interaction, which affects soil formation through freezing and thawing cycles.
- The topography, denoting where soil is located within the landscape, at the top, middle, or bottom of a slope, which has dramatic effects on the outcome of soil formation.
- Organisms, such as the dominant plant community and associated soil organisms that influence soil formation strongly by depositing OM and aggregating soil minerals.
- Time that has passed since the bedrock was laid down in relation to all of the other factors.

These factors combined explain the complex mix of characteristics that differentiate soil types. That soil types can vary considerably over short ranges illustrates the important role of the biota in soil formation because the other factors vary at larger scales both spatially and temporally.

Physical Components of Soil Systems

A typical soil is composed of both a mineral fraction and an organic fraction. These two fractions make up the soil solids, with the remaining soil volume composed of pore space, which at any given time is filled with some combination of air and water. When soil is saturated with water, all of the air in its pore spaces will have been displaced; conversely, desiccated soil has only air in the spaces between its soil solids.

The SOM content, the nature of the mineral fraction, and the relative proportions of air and water are critical factors affecting microbial activity and function. Soils with their pore space dominated by water are anaerobic. This condition will limit microbial activity to that of anaerobes and facultative anaerobes, i.e., organisms capable of metabolism in the absence of oxygen (O_2). The anaerobic process of fermentation is energetically less efficient than aerobic metabolism, and its end-products are generally organic acids and alcohols, which can be toxic to plants and many microbes. Hence, a soil with much of its pore space occupied by water much of the time will be a less productive soil, even though water is one of plants' critical needs.

A balance, where about half of the soil's pore space is occupied by air and half by water, is more supportive of both plant growth and microbial metabolism. Roots require O_2 in order to respire, and aerobes (microorganisms capable of aerobic respiration) can derive vastly more energy from this process than can be derived through fermentation or anaerobic respiration.

The nature of the mineral fraction determines the soil texture, content, and concentration of mineral elements as well as the presence of heavy metals, which can have some undesirable effects on plant and animal life. Phosphorus (P), potassium (K), and magnesium (Mg) are essential plant macronutrients derived from the soil mineral fraction. Hence, the productive capacity of any soil is very dependent on the composition of its mineral fraction.

Physical Properties and their Implications for Soil Biology

Other important soil physical properties include texture, bulk density, temperature, aggregation, and structure. Each has important effects on the composition and activity of soil biota.

Texture, which refers to the proportions of sand, silt, and clay in any given soil, will strongly affect the soil's water-holding capacity and its cation- and anion-exchange capacities. The ability of soil to retain water is important because microbes depend on soil water as a solvent for cell constituents and as a medium through which dissolved nutrients can move to their cell surface. Also, water is needed to facilitate the movement of flagellated bacteria, ciliated and flagellated protozoa, and nematodes. Texture thus directly influences biological activity in soil.

Bulk density refers to the weight of soil solids per unit volume of soil. Soils with a bulk density, 1 g cm23 are lighter or loose soils, likely to have good aeration and easy for roots to penetrate and for microbes to navigate. Soils with a bulk density.1 g cm23 are considered as increasingly heavier or compacted soils. As bulk density increases, soil porosity decreases, and air and water flows become restricted. This impedes soil drainage and root penetration. Such soils are often prone to waterlogging, creating anaerobic conditions.

Temperature will have varying effects on microbial activity depending on the respective organisms' range of tolerance. Psychrophilic organisms thrive in cold soil, at temperatures, 108C; mesophiles have their greatest rates of activity at temperatures between 10–308C; while thermophiles are more active at temperatures in excess of 408C. Soils in temperate regions experience prolonged periods annually at each of these temperature optima. This leads to marked seasonal shifts in microbial community composition throughout the year and to concomitant changes in the rates of SOM turnover and in the amounts of microbial biomass. Microbial communities in tropical soils also vary seasonally, but this is less determined by temperature.

Soil aggregation is the result of many interacting factors. In their model of soil aggregation, Tisdall and Oades described the process of aggregation as beginning with the interaction of clay platelets with one another at a scale of 0.2 mm. Microbial colonization of soil particles comes into play at a scale of 2 mm, an order of magnitude greater where bacterial and fungal metabolites serve to glue clay particles together. At a scale of 20 mm, fungal hyphal filaments and various polysaccharides produced by bacteria become the dominant aggregating factors. Then at a 200-mm scale, roots, and fungal hyphae bind these particles together. The resulting soil is a matrix of mineral particles bound together by biological materials at various nested scales to form macro aggregates at the 2-mm scale.

Soil structure describes the extent of micro- and macro aggregation of a soil. A well aggregated soil is more resistant to erosion from rain and wind. Also, it is generally well drained and more conducive for the growth of aerobic populations. It thus tends to be a more productive soil for plants and the soil biota. The process of aggregation as seen in the preceding discussion is the result of activities of plant roots and soil biota, creating intrinsic bonds between physical and biological characteristics of soil systems.

Influence of Soil Chemical Properties

Soil chemical properties strongly influence the activity of soil organisms, being at the same time themselves affected by such activity. The more important soil chemical properties affecting on biological activity are:

- pH, i.e., the acidity or alkalinity of a soil.
- Cation- and anion-exchange capacity.
- Mineral content and solubility.
- Buffering capacity.
- The concentration of nutrient elements in the soil.
- The concentration of O_2, carbon dioxide (CO_2), nitrogen (N_2), and other gases in the soil atmosphere.
- Soil water content.
- Salinity or sodicity.

Both plants and soil organisms have varying tolerances to extremes in soil pH. Most organisms prefer near-neutral pH values between 6 and 7.5. Many soil nutrients are most available for uptake by plant roots within this pH range. When soil is more acidic, the metal elements Fe, manganese (Mn), zinc (Zn), and copper (Cu) increase in solubility, while the solubility of most major nutrient elements — nitrogen (N), P, K, Ca, Mg, and sulfur (S) — decreases. The availability of N, K, S, and molybdenum (Mo) is unaffected at high pH; however, that of P, Ca, Mg, and boron (B) decreases above pH 8.0. In general, fungi and actinomycetes (bacteria that resemble fungi in their morphology and growth habits) appear to be relatively tolerant of both high and low pH, whereas many autotrophic and other heterotrophic bacteria are inhibited at low pH. Hence, in acidic soils, fungi and actinomycetes will tend to predominate. Organisms with greater limits of tolerance to changing abiotic conditions will have a competitive edge, which can affect the activity of others through substrate competition and thus inhibit their growth further.

Living organisms require a range of nutrient elements for their survival. Plants obtaintheir C (from CO_2), hydrogen (H_2) and oxygen (O_2) from the atmosphere, while the remaining elements must be derived from the soil solution. For most soil microbes, the situation is somewhat different as they derive their energy and cell biomass C mainly from decomposing plant and animal residues and from SOM. Notable exceptions include the cyanobacteria and other photosynthetic bacteria that fix CO_2 directly into cell biomass C using light energy, and the chemolithotrophic bacteria that use the bond energy in reduced compounds, such as NH_4, to generate reducing potential to fix CO_2 into cell biomass C chemosynthetically.

There are many pathways by which soil organisms obtain their energy, cell biomass C, and nutrients. Soil microbes obtain many of their other needed elements from the soil solution or soil minerals, which they solubilize to acquire the necessary nutrients, or from the soil atmosphere. Nitrogen is a special case. Almost 80% of the atmosphere is made up of nitrogen (N_2) gas. However, atmospheric N_2 is not available to plants until it has been reduced, either industrially, atmospherically, or through the process of biological nitrogen fixation (BNF). Many bacteria and cyanobacteria have the ability to fix N_2, but the most well-known are the rhizobia that fix atmospheric nitrogen in symbiosis with host legumes. Nitrogen-fixing bacteria, such as Azospirillum and Azotobacter, also form endophytic or associative relationships within or in close association with plant roots, and there are many free-living N_2 fixing bacterial species as well. Most soil fauna meet their energy, cell biomass C, and mineral nutrient requirements from consuming other organisms as either grazers or as predators.

The availability of mineral elements is not is the only important aspect; so are the relative proportions or ratios of mineral elements in relation to an organism's needs. A soil may be high in P, Mg, Ca, and S, for example, but if nitrogen availability is low, then the growth of soil organisms will be limited by the lack of this element. This concept is known as Liebig's "Law of the Minimum," where the growth of any organism is restricted by whatever nutrient element is in the shortest supply in its environment relative to its needs. This concept is important to bear in mind.

No matter how much of a given mineral nutrient is added to a soil, this will not improve crop yield or microbial growth if this is not a factor that is limiting production.

Adaptations to Stress

Given the high spatial variability in soil properties, the microorganisms that live in soil must be capable of rapidly adapting to continually changing surroundings. Soil organisms respond to stress by varying their use of O_2, by forming resting structures, by increasing intracellular solute concentrations, by producing polyols and heat-shock proteins, and by altering membrane structure, to name a few of the possible mechanisms.

Microorganisms vary in their need for or tolerance of O_2. We referred above to the two major groups in terms of their functional relationship to O_2: Aerobes and anaerobes.

Aerobes are species capable of growing at the O_2 concentration found in the atmosphere (21%), and they typically use O_2 as a terminal electron acceptor in the respiratory electron transport chain. There are three main types of aerobes: Obligate, facultative, and microaerophilic. Obligate aerobes require the presence of O_2 for their survival; their type of metabolism is aerobic respiration. While facultative aerobes do not require O_2, they grow much better if O_2 is present. These versatile bacteria have the capacity to respire either aerobically or anaerobically. Microaerophiles require O_2, but they can function at much lower levels than atmospheric concentrations. Their form of metabolism is aerobic respiration.

Anaerobes, on the other hand, do not or cannot use O_2 as a terminal electron acceptor. There are two basic types of anaerobes: aerotolerant anaerobes and obligate anaerobes. The first do not use O_2 for theirmetabolism, but they are not harmed by its presence. These organisms depend on a fermentative type of metabolism for their energy. Obligate or strict anaerobes, in contrast, are harmed by the presence of O_2. These organisms metabolize various substrates to derive energy either by fermentation or anaerobic respiration.

Facultative aerobes, microaerophiles, and aerotolerant anaerobes are better able to persist in the soil environment since they have the ability to adapt readily to the often rapid changes in O_2 availability that invariably occur in the soil. The capacity of facultative aerobes for use compounds other than O_2 as terminal electron acceptors in anaerobic respiration, for example, allows them to continue to respire C substrates and to generate the energy-storing molecule ATP via the electron transport chain when O_2 supply is reduced or cut. Nitrate (NO_3^-) and sulfate (SO_4^{2-}) are commonly used as alternative electron acceptors in anaerobic respiration.

The capacity to form spores or cysts is another type of adaptation that can enhance an organism's' persistence in soil during periods of low water availability. Bacterial endospores are very durable, thick-walled dehydrated bodies that are formed inside the bacterial cell. When released into the environment, they can survive extreme heat, desiccation, and exposure to toxic chemicals. Bacteria, such as Bacillus and Clostridium that form endospores, and actinomycetes and true fungi, that commonly reproduce by conidia and spores, are well represented in the soil community. Their capacity to form spores gives these species an obvious survival advantage in the soil environment. The much larger protozoa and nematodes which feed on bacteria and fungi can both form cysts or thick-walled resting structures that enable them to survive when conditions are not favorable for growth. Once conditions become favorable, such as after a rain or when prey populations increase, the cysts germinate and these protozoa and nematodes then resume feeding, growing, and reproducing.

Other adaptations also enhance the capacity for organisms to survive in the ever changing soil environment. Examples include producing polyols (alcohols with three or more hydroxyl groups) and heat-shock proteins; increasing intracellular solute concentrations; altering the membrane composition as seen in many Archaea (a prokaryotic lineage distinct from the Bacteria); and producing heat-stable proteins as seen in the thermophiles. In the last two decades, there has been a great increase in our knowledge of the survival strategies and mechanisms of soil biota which make possible the existence of the plethora of species that we are now coming to know, through molecular methods, are present in the soil.

Build it and they will Come

When the physical and chemical characteristics of a soil are within optimal ranges, biological activity generally follows suit. For example, if soil texture and structure allow for a good balance between adequate drainage vs. moisture retention with sufficient gas exchange, conditions will generally be conducive for microbial growth and activity. If the soil is compacted or water-saturated, it rapidly becomes anaerobic. Under such conditions, fermentative metabolism may predominate, and organic acids and alcohols are produced. Practices that improve SOM content, water-stable aggregation, and drainage, such as growing cover crops and retaining residues, applying compost, and reducing tillage all help promote abundant, active soil biological communities.

Classifying Organisms within the Soil Food Web

Soil Food Web as a System

When one thinks of any ecosystem, generally the first things that come to mind are the organisms — plants, animals, and microbes — that live within it and provide a variety of ecosystem services. In ecological terms, these are classified either as producers (plants, algae, and autotrophic bacteria) or consumers (herbivores, predators, and decomposers).

The primary producers, most often plants in terrestrial ecosystems, form the base of the food chain, or more accurately, the food web — a vast network of feeding interactions between and among organisms within the system. Primary producers capture energy from sunlight through the process of photosynthesis. This captured energy, stored in chemical bonds, provides the energy for most other organisms within the food web.

Trophic (feeding) interactions can be quite complex, especially below ground. Primary producers, generally plants, are consumed by herbivores, which are the primary consumers. Herbivores are in turn consumed by predators, which are considered secondary consumers within the system. Predators are then consumed by higher-order predators, the tertiary consumers within the system and on upwards.

Consumption is an energetically inefficient process. A rule of thumb is that only 10% of the energy contained at the first trophic level persists as usable energy at the next trophic level. Thus, up to 90% of the energy contained in primary producers, when consumed, becomes unavailable for metabolic work, being mostly lost from the system in the form of heat. This inefficiency of energy flow from one trophic level to the next has important consequences for the structure of ecosystems. The biomass that can be supported at any particular trophic level depends on the amount and availability of biomass in organisms at the trophic level immediately below it, upon which it feeds.

In above ground systems, the largest biomass will be that of the primary producers. As one moves to higher trophic levels in the food web, both the biomass and often the number of organisms that can be supported decrease. This leads to the concept of a pyramid of biomass, or a pyramid of energy. This shape suggests how the size of successive populations in any food web, i.e., their number and biomass, will decrease. Food webs will have, necessarily, a finite number of trophic levels as the total energy available for metabolic work at higher levels is consecutively dissipated as heat.

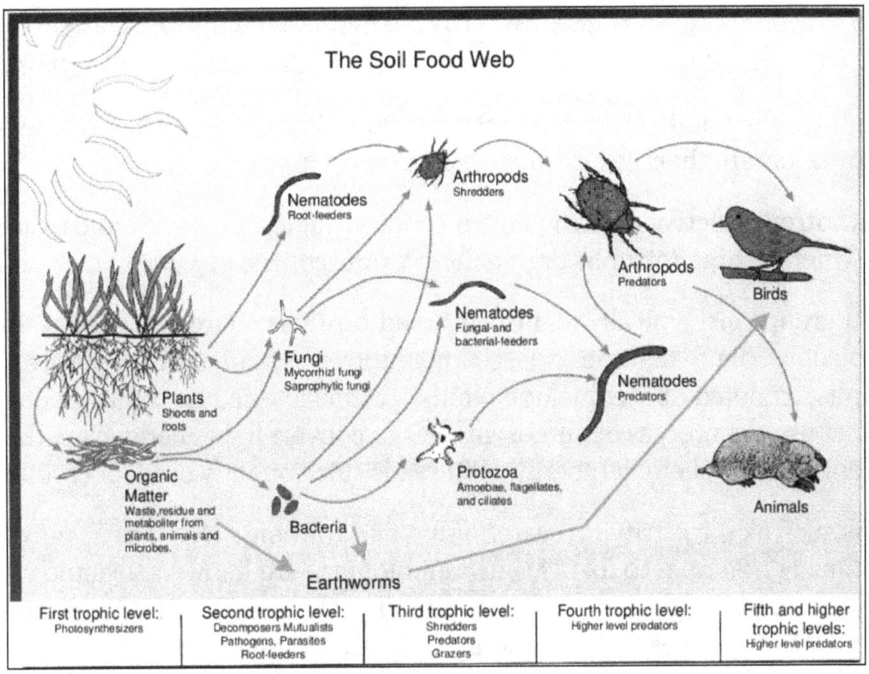

A simplified soil food web emphasizing trophic (feeding) relationships and functional roles of the soil biota.

Organisms in all ecosystems are dependent on a source of energy that can be captured to do metabolic work. Whether they capture it themselves through photo- or chemosynthesis or rely on preformed organic compounds, such as plant or animal tissue from other organisms, is a distinction that becomes very important when we consider the biota within an ecosystem's soil subsystem. The biological system beneath the soil surface operates on the same principles as those above ground, but with some distinct and important differences. The key difference is that primary production is extremely limited below ground since it is not continuously driven by abundant solar energy. This makes the whole subterranean subsystem energy-limited.

Root-derived soluble C compounds, sloughing of root cells, and root death below ground, plus litter and animal waste deposited above ground, are the primary sources of energy for the below ground community.

Energy and Carbon as Key Limiting Factors

The necessary goal for any organism is to obtain enough energy, cell biomass C, and mineral nutrients to produce the cellular constituents that are necessary for survival, growth, and reproduction. Metabolism refers to the biochemical processes occurring within living cells that make it possible for organisms to carry out what is necessary to maintain life. Microorganisms can be differentiated, and are categorized, based on three important metabolic requirements: (1) Their source of energy; (2) Their source of cell biomass C and (3) Their source of electrons or reducing equivalents.

- Phototrophs obtain energy from light, whereas chemotrophs obtain their energy from the chemical bonds in reduced organic or inorganic compounds.
- Autotrophs obtain their cell carbon from either CO_2 or HCO_3, whereas heterotrophs obtain their cell C from organic compounds.
- Lithotrophs derive electrons from reduced inorganic compounds such as NH_4 þ, whereas organotrophs derive them from reduced organic compounds.

Four main groups are typically identified based on their sources of energy and cell C: Photoautotrophs, photo heterotrophs, chemoautotrophs, and chemo heterotrophs. Photoautotrophs, as noted above, include plants, cyanobacteria, and other photosynthetic bacteria that use the process of photosynthesis to convert light energy from the sun into chemical energy. The chemical energy captured is subsequently used for carbon fixation.

Organisms such as the nitrifying bacteria that use ammonia (NH_4) as a source of energy and reducing potential to fix CO_2 into cell biomass are known as chemoautotrophs. Those bacteria and fungi, protozoa and soil fauna that rely on plant and animal residues and SOM as sources of both energy and cell biomass C are classified as chemo heterotrophs, or simply as heterotrophs. Photo heterotrophs are a small and unusual group of photosynthetic bacteria, the green no sulfur and purple no sulfur bacteria that use light as a source of energy and organic compounds as their source of cell C.

The activity of heterotrophic soil organisms depends on the availability of degradable organic C compounds. Since primary production below ground is limited by a lack of light, soil heterotrophs must depend on the activity and success of aboveground photoautotrophs, mainly plants, for their survival. In a healthy soil, heterotrophs meet their needs for energy and cell biomass C from the continuous addition of plant and animal residues, from the secretion of organic compounds by plant roots, and from the slow turnover of SOM, which includes the microbial biomass that continually dies off as new microorganisms come to life.

Factors affecting Microbial Community in Soil

Soil Moisture

Moisture is present in the form of film in soil pores. The amount of water increases with increase in porosity of soil. Pore-size depends on soil texture i.e. composition of sand, silt and clays. Moreover, soil moisture is affected through irrigation, drainage or management practices like tillage or crop rotation that enhance the intake and transmission of water by soil.

Organic and Inorganic Chemicals

The chemicals are very important for microorganisms as these provide nutrition for growth, activity and survival of microorganisms in ecologically deficient niches in soil. The chemical factors are gases, acids, micro- and macro-elements, clay minerals, etc. In the soil solution gases (oxygen, methane and carbon dioxide), and microorganisms are dissolved.

However, the dissolved components are in constantly shifting equilibrium with the solid phase, soil air, and moisture as well as with soil organisms and plant root activity. It has been found that low potassium and high nitrogen favour cotton wilt by Fusarium vasinfectum.

Soil-borne fungi are sensitive to pH. As a result of pH range for vigour and growth, they are more destructive at acid and neutral at alkaline conditions.

For example, Plasmodiophora brassicae favours best in acid soil, and the disease produced by it is uncommon or mild in soil of pH more than 7.5. Acidophilic natives of Trichoderma viride increased in soil on addition of sulphur, carbon disulphide, and methylbromide due to lowering down of pH to about 4.0.

Soil Organic Matter

Peat bogs.

The dead organic material of plant and animal origin serve as total soil organic matter which later is subjected to microbial colonization and decomposition. However, upon incorporation of green manures, crop residues, etc. in soil, the community size of microorganisms gets increased.

At the same time application of these organic matter alters the composition of soil microflora, microfauna, and relative dominance of antagonistic bacteria, actinomycetes, fungi, amoebae, etc.

Types of Vegetation and its Growth Stages

The dominance of one or the other groups is related to the type of vegetation and growth stages of a plant. Dubey and Dwivedi found an increased population of fungi in the non-rhizosphere and rhizosphere of soybean according to season and growth stages, respectively.

In the rhizosphere aspergilli, fusaria and penicillia were dominant in addition to the other fungal species. However, frequency of Macrophomina phaseolina and Neocosmospora vasinfecta increased on rhizoplane with onset of senescence.

This selective action of plants is attributed to microbial response either to specific root-exudates or chemical constituents of sloughed-off tissues that undergo decomposition. Moreover, Mueller determined the incidence of fungi and bacteria occurring in the roots of six soybean cultivars growing in fields cropped for 3 years either with corn or soybean.

Cropping history affected the recovery of M. phaseolina, Phomopsis spp. and Trichoderma spp. but not Fusarium spp. or Gliocladium roseum. Recovery of Trichoderma spp. was greater following com than following soybean. After death of the plant soil saprophytes colonize rapidly, thus total spectrum of microflora in the rhizosphere is changed.

Different Seasons

The amount of plant available nutrients is governed by the number and activity of microorganisms. They remain in constant dynamic state in soil where microbial community is greatly influenced by physicochemical and biological factors. Changes in microbial community are known in soils of tropical, sub-tropical and temperate regions.

Physical Factors affecting Soil Organisms

There are many physical factors which affect the activity of the soil biota. The main ones are:

- Temperature.
- pH.
- Moisture.

- Soil mineralogy.
- Light.

Algae.

Temperature

Temperature directly affects the activity of the soil biota by determining the rate of physiological activity such as enzyme activity and indirectly by affecting physico-chemical properties such as diffusion & solubility of nutrients, mineral weathering and evaporation rates and so on.

Within defined limits biological activity increases with increasing temperature. Organisms have a specific range of temperature at which their biological activity operates. For common soil organisms the temperature range at which they can be active ranges from about 0 °C to about 60 °C although no single species is likely to be active throughout the entire range.This temperature range is largely determined by the temperature at which soil can reach although organisms may be able to survive outside the soils' temperature range or produce survival structures to allow them to survive under adverse conditions.

There are extreme environments such as volcanic hot springs & vents which can reach considerable higher temperature. It was once thought that the highest temperature that could be survived by microorganisms was around 400 °C but this has been reduced to around 121 °C.

Within an organisms' temperature range, there is a temperature optimum at which biological function performs best. Beyond this value, cellular processes do not work so efficiently and as the temperature increases away from their upper limit there are irreversible changes to the cell properties leading to cell death. Those organisms which have a high temperature limit would tend not to be so active at the lower extreme compared to organisms more suited to the lower limit.

Soil pH

In a similar manner to temperature, organisms have a range of pH at which they are active. Within certain limits, organisms can tolerate extremes but this normally requires the cell to use energy in maintaining the correct internal cellular pH (pH 7.0). A few organisms (bacteria & archaea) can tolerate very extreme pH values such as pH 1 or pH 11 but these are extreme conditions are not found in agricultural soils. Larger soil animals may be more sensitive to pH than microorganisms. Earthworms being sensitive to low pH are not active in forest or peatland systems (pH <5) their function is replaced by other organisms (enchytraeids in forests) or the soil forms in a different manner (peatlands).

pH directly affects the solubility of elements. At acidic pH, aluminium becomes more soluble and hence more available to the organisms with increased toxicity. Essential minerals can become unavailable at extremes of pH. For example, phosphorous and managanese become increasingly unavailable at high pH values.

Soil Moisture

Soil moisture affects the soil biota in two ways. Biologically water is essential for life and for enzyme activity and metabolism and, is a solvent for biological nutrients and other chemicals. Physically, soil moisture affects soil temperature (water is good conductor of heat) and soil aeration. The degree at which soil pores are filled affects the movement and predation of microorganisms in soil. In very dry soil, plants may not be able to extract sufficient water through the roots because of the energy it takes to remove water from the small pores. This is known as the permanent wilting point, beyond which the plant cannot recover. Conversely, under wet conditions, oxygen does not diffuse through the soil as readily so the levels available to organism may become depleted leading to anaerobic conditions. Fungi tend to be more resistant to water stress than bacteria although under extreme conditions, some bacteria can form resistant endospores which allow the organism to survive until more suitable conditions arise. Chemicals may be produced by the cell to allow survival under drought. Actinomyces which can tolerate drier conditions than fungi can protect themselves from drought by synthesising the amino acid proline.

Soil Mineralogy

Soil minerals are weathered by bacteria and fungi either as a direct source of energy for the soil biota (chemoautotrophs) or a consequence of the production of acids such as sulphuric acid. As well as weathering minerals, bacteria have been shown to form minerals. Depending on the pH, some nutrients particularly ammonium may become bound up with the clay an unavailable. Clays have a number of sites on their surface which can bind enzymes and whole organisms thereby affecting the movement of cells through soil and the breakdown of organic matter.

Light

In soil, light directly affects those organisms on or just below the surface and indirectly by heating the soil surface. Phototrophs, such as plants, algae and cyanobacteria, use the energy from sunlight to synthesis carbohydrates. In plants, some of this material (the photosynthate) finds its way into the soil biota via the roots and from leaf fall. With algae and cyanobacteria, the organic matter is directly inputted into soil either from release of material from the cell or when the cell dies.

Parts of the light spectrum are more damaging to organisms than others. Ultraviolet light can damage DNA which induces mutations in the organisms resulting in death of the organism or biochemical changes through for example changes of enzymes and metabolic pathways. Photochemical degradation will lead to structural changes which organic molecules making them more or less easy to be degraded.

Relationship of Agronomic Practices to Soil Nitrogen Dynamics

Agronomic Practices Reflect the History of Managing Soil N Dynamics

Nitrogen (N) is the most important plant mineral nutrient. It was first discovered in the late eighteenth century, and N's role in improving crop production was widely recognized by the mid-nineteenth century. Long before these discoveries, ancient farmers often unknowingly employed agronomic practices that resulted in managing soil N availability, thereby helping to ensure the human food supply and nutrition. There were two major sources of N in agroecosystems before synthetic N fertilizers—soil N and legume-based biological N fixation. Ancient farmers constructively developed tillage schemes and rotated nonlegume and legume crops to manage both N sources for millennia. Because the appearance of commercial synthetic N fertilizers in the early twentieth century brought significant changes to traditional agronomic practices.

Plow tillage is a form of soil N management. Much of the soil N is in complex organic forms, such as decomposing plant and animal residues. Most plants can only take up inorganic N (NH_4^+ and NO_3^-) although the basic amino acids are absorbed by some plant species (e.g., Picea abies). Inorganic N and basic amino acids in soil are mainly derived from N mineralization processes. Tillage practices can promote mineralization because disturbance exposes naturally protected (i.e., aggregate-protected) soil organic matter (SOM) to microbes, enhancing microbial activity and N mineralization. Therefore, plow tillage was considered a great agricultural advance and, from the archeological evidence, has had a very long history. Foot plows also called "digging sticks" are shown in

Egyptian tomb paintings. A wooden model of oxen and plow was found in an Egyptian tomb dating from 2000 BCE. In Asia, one of the oldest existing Chinese books titled "Lü Shi Chun Qiu" or "The Annals of Lu Buwei" demonstrated the details of when and how to till according to soil and weather conditions and served as an early example of a practical farming guide.

Rotation can also be a tool to manage soil N through legume bio-fixation of N, depending on the crop species. Monocropping, especially with nonlegumes and heavy-nutrient-using crops (e.g., tobacco and corn) can deplete soil N. Rotation practices, even simple fallow, help to restore soil N. This practice was evident in early Roman times. One of the Rome's greatest poets, Virgil, wrote in his poem Georgics "For the field is drained by flax-harvest and wheat-harvest, drained by the slumber-steeped poppy of Lethe, but yet rotation lightens the labour." This emphasizes that fallow was necessary to rotate with those crops requiring more nutrients. On the other hand, rotations that include a legume crop can bring biological N fixation into agricultural production systems. Although ancient farmers knew nothing of the biological N fixation process and nothing about the importance of mineral N to plant growth, they intentionally included legume crops into crop sequences. This was evident in the Pliny the Elder's book on natural history that mentioned several legume successions as alternatives to conditions that forbade fallowing.

Synthetic fertilizer N application in agricultural production has a relatively short history compared to tillage and rotation practices because knowledge regarding N in plant nutrition and N synthesis techniques is recent. In 1836, Jean-Baptiste Boussingault investigated manure, crop rotation, and N sources and for the first time concluded that N was a major component of plants and that the nutritional value of fertilizer was proportional to its N content. However, ammonia could not be easily synthesized from constituent elements until 1908, when the Haber-Bosch process was developed. After that, synthetic fertilizer N started to play a greater role in agricultural production, helping to improve global food security.

Influence of Synthetic Fertilizer N on Traditional Agronomic Practices

The appearance of synthetic fertilizer N brought a huge increase in the global food supply. Erisman et al. estimated that around 50% of the world population's food requirements are currently met by using synthetic fertilizer N. However, synthetic fertilizer N fundamentally disturbed the soil N cycling balance in agro-ecosystems and brought significant changes to traditional agronomic practices. Our unpublished data show that synthetic fertilizer N can promote or prime soil N mineralization depending on the indigenous SOM level and the amount of synthetic N.

Synthetic fertilizer N played a role in developing modern no-tillage farming. Agriculture derives numerous benefits from no-tillage, including fuel and labor savings, increased soil C stocks and erosion resistance. But few people recognized the fertilizer N contribution to no-tillage until early Kentucky no-tillage × N fertility trials revealed its importance. No-tillage without N fertilizer significantly lowered yield compared to conventional tillage without N fertilizer. However, no-tillage with N fertilizer produced yields comparable to those of conventional tillage with fertilizer N. From this perspective, one can speculate that added fertilizer N compensated for reduced soil N mineralization in no-tillage. Other factors, including herbicide and equipment development, also made no-tillage farming feasible in Kentucky and the rest of the Southeast and Mid-Atlantic States in the USA, beginning in the 1960s. At the time, the move away from tillage was viewed with much skepticism, but eventually no-tillage was accepted as a revolution in farming. By 2009, approximately 36% of U.S. cropland, planted to eight major crops, was in no-tillage soil management.

Although ancient farmers knew nothing of biological N fixation, legume crops had been an important cropping system component worldwide before synthetic N became available. However, crop rotation was discouraged during the Green Revolution, partially because pest control benefits from crop rotation could be replaced by chemical crop protectants. Also, the N credits from biological N fixation could be easily replaced by synthetic fertilizer N. However, soon after the height of the Green Revolution, many studies reported that no amount of chemical fertilizer or pesticide could fully compensate for crop rotation benefits. Rotation systems then came back into fashion. Currently, 80% of all corn, soybean, and wheat planted acres in the United States are in rotation.

Systematic Understanding of Agronomic Practices and Soil N Dynamics

Managing N dynamics is one of the central reasons farmers developed and implemented specific agronomic practices. Furthermore, in the last few decades, new knowledge indicates how transient N can have negative impacts on global environments and human health. A systematic understanding of "How does soil and crop sequence management influence nitrogen dynamics?" will significantly influence agronomic practice development but also has global meaning for the quality of human life. The aim of optimal agricultural N management is to enhance net N mineralization at times when crops need N, to synchronize soil N mineralization with crop N uptake, and to minimize N loss. To systematically understand this topic, three sequential steps need clarification:

- How do agronomic practices affect soil organic matter pools?
- How do soil organic matter pools contribute to soil N availability?
- How do agronomic practices influence crop N uptake capacity?

Soil organic and crop residue N pools provide the organic N for N mineralization. This microbial process, primarily heterotrophic, also requires soil organic C (SOC) as an energy source. Thus, to understand how soil and crop management affect mineralized soil N, it is critical to first evaluate whether and how tillage, rotation, and fertilizer N application affect SOC and N sequestration. Soil organic matter sequestration has been reported to be linked with soil aggregate formation. The dominant concept that explains SOC and N sequestration is based on the aggregation-SOM model. Zou et al. reported that using NT and rotation practices in burley tobacco production maintained desirable soil physical and chemical properties by macroaggregate stabilization, which led to conserving SOC and TSN stocks. The basic idea is that soil organic matter functions as a nucleus/binding agent for aggregate formation. Aggregates are important reservoirs of SOC and N that are protected from microbial access and less subject to physical, chemical, microbial, and enzymatic degradation.

Appropriate and precise estimation of soil N mineralization has been a challenge since the early 1900s. Temporal and spatial variability are large because this process is determined by internal soil factors (e.g., SOM level, labile C and N pools, soil microbial community) and external environment factors (e.g., temperature, precipitation and aeration). Agronomic management, such as plant species and N fertilizer application, may also affect N mineralization. With current technologies, it is impossible to predict N mineralization by taking these factors into consideration simultaneously. Instead of being a measure of available N supply, N mineralization estimates by current methods should be considered an index of N availability.

Isotopic tracers and incubation methods are the two main approaches used to estimate N mineralization. The isotopic tracer method can measure gross N mineralization, but isotope methods are expensive and can also have methodological problems with mineralization rate estimates and other assumption violations. Although incubation methods can only measure net soil N mineralization (net soil N mineralization = gross N mineralization–N immobilization), incubation can fairly estimate the available N pool, which has a practical value for efficient N management in agroecosystems. Therefore, long-term biological mineralization has been considered the most suitable soil N availability index and is often used to validate other indices derived from more rapid chemical or biological assays. There are, however, many variations to incubation methods, including environment, sample pretreatment, and incubation time, and each variation has advantages and disadvantages. To use incubation to meet research objectives, assumptions, benefits, and liabilities of each variation should be considered.

An experimentally derived N availability index might not necessarily reflect total crop N uptake. Besides the amount of available soil N, crop N accumulation also depends on N uptake capacity. Crop N uptake capacity might be determined by either/both genetic and environmental controls. Genetics can control crop growth rate and biomass accumulation, which would result in different N demands at different growth stages. Crop species have different root architectures, mostly controlled by genetics. However, roots, the dominant nutrient uptake organ directly exposed to the soil, interact with a wide array of soil physical, chemical, and biological factors that vary in time and space.

To understand the impact of agronomic management practices on crop N uptake or yield, both soil N availability and root architecture must be considered.

Mechanisms and Effect of Agronomic Practices on Soil C and N Sequestration

The link between SOC and total soil N (STN) decomposition and stabilization and soil aggregate dynamics has been developed, recognized, and intensively studied since the 1900s. Soil organic C and N dynamics are important to agricultural production because these affect soil nutrient cycling and plant productivity. The C and N dynamics are also important to the environment because they can affect greenhouse gas emissions and water quality. These processes happen in a heterogeneous soil matrix and have multiple interactions with soil biota. The task of elucidation is complex. Aggregate-SOM models can explain some of these complexities. Aggregates not only physically protect SOC and SON, but also influence soil microbe community structure, limit oxygen diffusion, regulate water flow, determine nutrient adsorption and desorption, and reduce surface runoff and erosion. All these processes have fundamental effects on soil C and N sequestration and stabilization.

More current studies to understand the impact of agronomic practices on soil C and N sequestration have been based on the aggregate hierarchy concept proposed and developed by Tisdall and Oades. To apply the theoretical aggregate-SOM models, the first consideration is the physical separation of soil into different aggregate size classes. Two main methods to separate soil aggregates are widely used by researchers: dry and wet sieving. The disruption of aggregates is mainly due to slaking and microcracking when the soil is initially dry. Dry sieving of air-dried samples is used to characterize the aggregate size distribution with minimum destruction. Wet sieving is used to simulate microcracking and slaking. Water-stable aggregate stability from wet sieving procedures was reported to be closely correlated with SOM stabilization because SOM can act as a transient binding agent and has served as an effective early indicator of soil C change in numerous studies. The wet sieving procedure has been frequently used to evaluate the agronomic practice effects on both SOM sequestration and soil structural stability. In the wet sieving procedure, sample pretreatment is important. The rewetting pretreatments for soils can cause different results when comparing soils and management history treatments. Cambardella and Elliott showed that capillary-wetted soils retained more macroaggregates (>250 μm) than slaked soils. Bissonnais demonstrated that the different aggregate breakdown methods and frequency of crusting soil samples can dramatically affect soil aggregate stability within the same soil management system. Adopting minimum breakdown aggregates in the sieving procedure keeps comparisons between treatments relative to the natural field conditions.

The effect of agronomic practices (including tillage, rotation, and fertilizer N application) on SOC and STN, according to aggregate-SOM models, has been studied intensively in grass and grain crop production systems, but not in leaf harvest crop production systems. In these studies, no-tillage increased or maintained SOC and STN compared to conventional tillage. With aggregate separation, conventional tillage can increase large aggregate turnover rate, diminishing the macroaggregate proportion and SOC and STN concentrations. In contrast, no-tillage increases macroaggregates and SOC and STN accumulation.

Most studies show that rotation increases SOC and STN sequestration, compared to monocropping. Crops in rotation schemes have different impacts on SOM stabilization, depending on the quantity and quality of crop residues. Wright and Hons found that crop residue production was similar among wheat, sorghum, and soybean fields, but the wheat field had significantly higher SOC and STN in surface soil than the other two fields, which indicates that the higher C:N ratio in wheat residue can play a role in SOM stabilization. Kong et al. reported that the quantity of crop residue/carbon production had a linear relationship with SOC sequestration in sustainable cropping systems. Therefore, when evaluating crop rotation schemes on SOM sequestration, examining crop residue quantity and quality is important.

Studies on the effect of fertilizer N application on SOM sequestration have produced the most controversial results. Some studies report that fertilizer N application increases SOM because higher fertilizer N input causes more crop residue to be returned to soil. Mulvaney et al. reported that fertilizer N application decreased soil N in the long-term Morrow plot study and argued that synthetic N application enhanced soil microbial decomposition due to the decreasing C:N ratio. Others have found no effect of N fertilizer application on SOM sequestration.

Methodologies of Soil N Mineralization Measurement

There are many different methods available for long-term aerobic incubation, in laboratory and field, depending on soil sample pretreatment and other incubation conditions.

Laboratory Incubation Methods

Most aerobic laboratory incubation methods have common features, including maintenance of optimal soil water status (typically with 60% water-filled pore space), constant temperature (commonly 25, 30, or 35°C), and periodic sampling to estimate N mineralization rates. Although there have been several standardized protocols, there is significant variation in aerobic incubation details.

Leaching versus Non-leaching Processes

In early studies with long-term N mineralization incubation, samples were usually incubated continuously in a container without periodic leaching of the accumulated inorganic N. The merit of this method was convenience, but there could be cumulative inhibitory effects, such as pH decline, on mineralization during the incubation. Thus, nonleaching approaches were not recommended for long incubation periods. Stanford and Hanway proposed a periodic leaching approach during incubation. Briefly, 0.01 M $CaCl_2$ was used to leach mineralized N from the sample at the end of each incubation period. The merit of leaching would be avoidance of accumulation of unspecified toxins. While being a time-consuming and apparatus-requiring process, there was also an additional technical concern with potential leaching of soluble organic N during the incubation.

Excluded Crop Residue versus Included Crop Residue

Crop residue can contribute to the soil inorganic N pool by N mineralization or immobilization, depending on the residue C:N ratio. Most laboratory incubation methods exclude such contributions by discarding visible pieces of residue in the pretreatment sieving process. Some laboratory methods cut entrained residues into pieces that are mixed with soil for incubation. Certainly, discarding large portions of residue might influence estimates of the N credit from the previous crop because soil fertility guidelines usually recommend a different fertilizer N rate for the current crop that depends on the previous crop.

Field-moist Soil Sample versus Dried and Ground Soil Sample

Using dried and ground soil is convenient for a large amount of soil samples that require time to process or for cooperative projects where soil samples come from multiple locations at different times. However, several days are needed to rewet soil for preincubation, which also causes an N mineralization flush during the first weeks of incubation. Numerous studies report that sample sieving and drying-rewetting causes rapid microbial death and enhances microbial respiration and activity, producing an N mineralization bloom. Using field-moist samples might cause less physical damage during preincubation protocols and cause a better transition from field to lab conditions than dried and ground soil samples. However, field-moist soil samples intended for incubation need to be gently crushed through the sieve (usually 2–4 mm) immediately after sample collection.

Homogenized Soil versus Undisturbed Soil Cores

Most laboratory incubation methods utilize a homogenized sample created by sieving. However, there are reports that homogenized samples do not well represent the effects of field soil tillage. Laboratory soil should have a physical structure similar to that of the field environment the sample represents, but sieving artificially "tills" soil from undisturbed/no-tillage environments. This can expose aggregate-protected SOM and enhance the microbial activity, over-estimating the N mineralization. Undisturbed cores may be a better option for laboratory incubations intended to differentiate the impact of tillage on N mineralization.

Constant Temperature versus Variable Temperature

Most laboratory incubation methods use a constant temperature, which does not reflect temperature fluctuation in field conditions. Carpenter-Boggs et al. proposed a variable temperature method for laboratory incubation in which soil samples are incubated in a variable temperature incubator (VTI) that mimicked field soil temperatures under a growing corn canopy. They reported that the VTI technique provided a lower sample variance and a smaller initial flush of N mineralization than constant incubation temperature (35 °C).

Field (In-situ) Incubation Methods

Due to the uncertainty about the extrapolation of laboratory N mineralization values in the field, estimating N mineralization from SOM and crop residues in field conditions would be a compelling research topic for investigators because more efficient N fertilization practices could be hastened if a reliable in-situ N mineralization method was developed. So far, there have been three dominant in-situ research techniques: buried polyethylene bags, covered cylinders, or resin-trap core methods.

Buried Polyethylene Bag Method

The buried polyethylene bag method for in-situ N mineralization was proposed by Eno. The main driving force behind this technical development was the realization that soil temperature variance results in considerable changes in the soil NO_3^- production rate. In that preliminary laboratory study, soil in sealed polyethylene bags had an equal nitrification rate compared to soil contained in ventilated bottles. Polyethylene is permeable to oxygen and carbon dioxide, but no NO_3^- diffused through the polyethylene bag during the 24-week incubation period. The preliminary results and polyethylene characteristics mean this technique has the potential to estimate aerobic in-situ soil N mineralization.

Although this technique mimics field temperature conditions at a low cost, the technique does not reflect transient field moisture conditions. Elevated NO_3^- and carbon dioxide concentrations inside the bags may promote denitrification. Physical damage to the bags by insects or plant roots may result in loss of mineralized N into the field soil via diffusion and mass flow. Another major limitation of this technique is the inevitable disturbance of soil, which does not allow a valid comparison of tillage effects on N mineralization in field conditions.

Covered Cylinder Method

The covered cylinder method was developed as a more durable alternative to the buried bag. This technique allows incubation of intact soil cores. Covered cylinders are usually constructed from PVC or metal pipes that are capped to exclude rainfall, which is also assumed to stop inorganic N leaching. Although the tubes are open at the bottom, aeration is less than that in field soil, which might result in higher denitrification potential. Modifications such as using gas permeable caps or perforations in the tube sidewall were often added to promote air exchange and reduce denitrification potential. However, sidewall aeration holes could potentially allow mineralized N loss. Water may enter the soil tubes through aeration holes, causing N leaching at the bottom of the soil column. Furthermore, plant roots may potentially grow into the soil column via aeration holes or the open bottom, absorbing mineralized soil from the tubes. Another major limitation of this technique is that the soil in the tube usually has a lower soil moisture content than the surrounding field.

The basic principle of the covered cylinder method was to limit N leaching by sheltering incubating soil from precipitation. Based on the same principle, there was another in-situ method called the "rain shelter", which simply used a shelter over the sampled area to prevent leaching. Except for considerations regarding the quality and durability of the rain shelter and surface water run-on during intense rainfall, the major drawback of this technique is a lack of ability to reflect field soil moisture fluctuations.

Resin-trap Soil Core Method

Buried polyethylene bags and covered cylinder methods can capture variation in field temperature, while failing to reflect moisture and aeration conditions, which are reported to play a large role in soil N mineralization. An alternative in-situ method was proposed that employs ion exchange resins to capture mineralized N leaching from undisturbed soil cylinders. The major modification of this technique is an open cylinder top, which allows precipitation and air to freely enter the intact soil column, and a resin trap at the bottom to capture inorganic N that might otherwise leach from the tube. There are some concerns about whether the soil tube causes abiotic differences between soil in the tube and the surrounding field soil. Wienhold et al. reported that soil inside the cylinders was slightly wetter and warmer than adjacent soil, which likely increases soil N mineralization. They pointed out that the magnitude of change in soil N mineralization was likely much less than the normally observed field core-to-core variation. This method was found to better track true field conditions and has the potential to become a standard procedure.

The drawback with intact cores and resin bags is a large resource demand. This technique requires preliminary studies to ensure leached ions are efficiently trapped under field conditions. Resin duality, adsorption capacity, and bypass flow are all factors that can potentially influence resin effectiveness in capturing leached N. The extraction of adsorbed N from the resin is also time-consuming. Kolberg et al. reported that five extractions with KCl were required to completely release adsorbed N.

Other Modifications to In-situ Incubation Methods

Except for the major design developments mentioned above, some minor modifications to in-situ incubation methods have been suggested. Hatch et al. proposed a method to combine the soil core with acetylene inhibition, which would limit N loss by nitrification due to uncontrolled soil in-situ incubation conditions. The big concern with this modification is that the tube must be sealed at the top, causing a loss in practical application to the field environment if rainfall is a concern. Given consideration on different drainage characteristics in resin-trap soil cores relative to the surrounding soil, Hanselman et al. developed a "new" type of resin-trap soil core method in which resin is mixed with soil to create an artificial uniform soil column. This method is impractical when undisturbed soil structure is a research concern, as in a comparison of conventional and conservation tillage.

Method Selection

Each method, including laboratory and in-situ methods, has unique assumptions, advantages, and disadvantages. There is no standard method that will work for every situation. The selection of method depends on the nature of the study, available resources, and site-specific factors. Although laboratory methods might not reflect natural field conditions, they can provide reasonable relative values to estimate differences due to soil type and certain management practices. Zou et al. reported that discarding plant residue in the laboratory incubation method neglects the potential effect of plant residue on soil N mineralization. The primary merit to field incubation is a more practical estimation of N mineralization, which might be more useful in management decision making. However, the substantial time and apparatus requirement for the in-situ incubation methods must be considered. Zou et al. reported that soil C and N fractions contribute variably to predict soil N mineralization in different rotation systems, but SOC (which can be calculated from soil organic matter, a common index in the routine test package of many soil testing laboratories) was the best overall NSNM predictor in their study. The principle is that both biotic and abiotic factors control the soil N mineralization process. Knowing the advantage and disadvantage of each method can help the investigator choose the best method while reducing misinterpretation.

Influence of Agronomic Practices on Root Architecture

Plant roots are a fundamental component of terrestrial ecosystems and function to maintain nutrient and water supply to the plant. Although root system architecture is controlled mainly by genetic factors, plant root systems exhibit high developmental plasticity. This plasticity is possible because root development results from continuous propagation of new meristems. In a heterogeneous soil matrix, a wide array of physical, chemical, and biological factors can affect the initiation and activity of root meristems. Previous studies have reported that certain crop root traits enhance productivity in resource-limited environments due to improved nutrient and water scavenging abilities. Agronomic practices can influence crop nutrient uptake capacity by affecting the root growth environment.

Tillage affects root growth mainly by changing soil structure, strength, and penetration resistance. Any particular root increases its length through primary growth when cells of the meristem divide, elongate, and push the root tip forward through surrounding materials. Turgor pressure in the elongating cells is the driving force and must be sufficient to overcome cell wall constraints and other additional constraints imposed by the surrounding environment. Compared to conventional plow tillage, numerous studies on grain crops report that no-tillage increases mechanical impedance, which can result in reduced root length density, root surface density, and lower biomass production. Similar results were found in a no-tillage burley tobacco study. Furthermore, greater mechanical impedance with no-tillage not only restricts root growth but also changes root morphology, restricting main root axis elongation, stimulating lateral root branching and root thickening.

Nutrient supply and distribution (or fertilizer application) can affect root system architecture mainly by signaling. Typically, roots proliferate in volumes where nutrients are most concentrated. However, the mechanisms of plant root response to the different nutrient elements might be controlled by different pathways and signals.

There have been few studies on the effect of crop rotation on plant root architecture. Given the basic factors controlling root development, the hypothesis is crop rotation may differentially influence root architecture compared to monocropping systems, if rotated with residue-rich or deep-rooted crops that can increase SOM levels and soil structure. In this case, rotation affects root proliferation by changing soil structure in a manner similar to that observed with no-tillage. If rotation involves legumes, more N nutrition is provided than found with monocropping. In this case, rotation could affect root architecture by changing soil nutrient supply in a manner similar to that found with fertilizer application.

The effects of agronomic practices on crop N uptake not only affect SOM sequestration and soil N mineralization but also alter the soil environment for plant root proliferation. Similarly, in a paper titled "A New Worldview of Soils", soil productivity is broadly defined as the soil's unique ability to supply water, nutrients, air, and heat, among other life-sustaining resources, adjusting that supply to the demands of plants and microbes. Soil resources fall into two main components: (a) Nutrients and moisture and (b) An environment suited for root growth and microbial activity.

References

- Soil-Environment: microbewiki.kenyon.edu, Retrieved 19 June, 2020
- The-Soil-Habitat-and-Soil-Ecology-279398426: researchgate.net, Retrieved 28 April, 2020
- 5-factors-affecting-microbial-community-in-soil-microbiology, soil-microbiology-66660: biologydiscussion.com, Retrieved 01 February, 2020
- Physical-factors-affecting-soil-organisms, soil-biodiversity-soil-organisms: fao.org, Retrieved 16 April, 2020
- Relationship-of-agronomic-practices-to-soil-nitrogen-dynamics, soil-productivity-enhancement: intechopen.com, Retrieved 08 July, 2020

The Rhizosphere 5

- **Rhizobacteria**
- **Rhizosphere Interactions**
- **Root Microbiome**
- **Rhizosphere Microorganisms**

The rhizosphere is the narrow region of soil around the roots where microbial biomass is affected by the presence of soil microbes, plant roots and their secretions. Some of the rhizosphere organisms are oomycetes, bacteria, pathogenic fungi, and human pathogens. All the topics associated with the rhizosphere have been carefully analyzed in this chapter.

The rhizosphere is a dynamic region governed by complex interactions between plants and the organisms that are in close association with the root. The composition and pattern of root exudates affect microbial activity and population numbers, which in turn have an impact on the nematodes and microarthropods that share this environment. Beneficial or harmful relationships exist between rhizosphere organisms and plants, which ultimately affect root function and plant growth. In addition, the rhizosphere may include organisms that do not directly benefit or harm plants but clearly influence plant growth and productivity. A better understanding of the soil–root and soil–seed interface is needed to manage microorganisms, increase plant growth, and reduce the impact of plant production and agriculture on the environment. The benefits of studying the rhizosphere include the use of plant growth-promoting organisms and the suppression of plant diseases and weeds using biocontrol agents. Rhizosphere organisms can also be used to enhance the formation of stable soil aggregates and as bioremediation agents of contaminated soils. With greater understanding of the ecology and biota in the rhizosphere, this zone of increased nutrients, biotic activity, and interactions can be manipulated to improve plant productivity and environmental quality.

The rhizosphere is a nutrient-rich region of the soil immediately surrounding the plant root. This region is highly dynamic and supports a dense and diverse fauna. Despite the challenges associated with studying ecological interactions in a soil matrix, researchers are beginning to understand the complex ecological interactions occurring in the rhizosphere. Chemical communication plays an integral role in the ecology of the rhizosphere, and new functions for intra- and interspecific signals continue to surface. Much of the knowledge on rhizosphere biology has been revealed by agricultural researchers, who have studied many of the positive and negative relationships between plants and soil microbes. Though detailed descriptions exist for many rhizosphere interactions, the complex and cryptic nature of the rhizosphere will continue to challenge scientists interested in the ecology of the plant–soil interface and its associate biota.

Rhizosphere Microzone

The rhizosphere, first described in 1904 by Lorentz Hiltner, has been the focus of intensive research for many years because of its importance in plant nutrition and pathogenesis. More recently, research on the rhizosphere has been directed toward its influence on controlling the persistence, mobility, and bioavailability of contaminants in soils.

The rhizosphere generally refers to the portion of soil found adjacent to the roots of living plants. The rhizosphere is subject to the influence of chemicals excreted by roots of living plants and the microbial community in this microzone. Its domain varies for different plant species and for age and morphology of roots. Depending on plant species, the width of the rhizosphere zone has been shown to extend from 2 to 80 mm away from the root surface. The soil in the rhizosphere supports a typically diverse and densely populated microbial community and is subject to chemical transformations caused by the presence of root exudates and metabolites of microbial degradation.

The rhizosphere microzone is distinguished from the bulk soil zone, more commonly known as the edaphosphere, by enhanced microbial activity and increased concentration of root exudates. Nevertheless, it is rather difficult to separate this zone physically from the root surface or rhizoplane. The rhizosphere effect is expressed quantitatively as the ratio of the number or activity of microorganisms or level of root exudates in rhizosphere soil (R) to that in the edaphosphere soil (E), i.e., the R/E ratio. The R/E ratio for microorganisms and root exudates is often found to range, respectively, from 2 to 20 and from 5 to 100, indicating enhanced microbial activity in the rhizosphere.

Root exudates are known as one of the most important factors affecting microbial growth in the rhizosphere. For example, cell numbers are several orders higher in the root zone than in the background soil lacking plants. The microbial community is more diverse, active, and synergistic than in nonrhizosphere soil. Root exudates can also selectively influence the growth of bacteria and fungi that colonize the rhizosphere by serving as selective growth substrates for soil microorganisms. For example, Pseudomonad species are particularly stimulated by enhanced carboxylic acid production from root exudation. Rhizosphere microbial communities can also vary in structure and species composition in different root locations or in response to other environmental factors, including soil type, plant species, or nutritional status.

Rhizobacteria

Rhizobacteria are root-associated bacteria that form symbiotic relationships with many plants. Though parasitic varieties of rhizobacteria exist, the term usually refers to bacteria that form a relationship beneficial for both parties (mutualism). They are an important group of microorganisms used in biofertilizer. Biofertilization accounts for about 65% of the nitrogen supply to crops worldwide. Rhizobacteria are often referred to as plant growth-promoting rhizobacteria, or PGPRs. The term PGPRs was first used by Joseph W. Kloepper in the late 1970s and has become commonly used in scientific literature. PGPRs have different relationships with different species of host plants. The two major classes of relationships are rhizospheric and endophytic. Rhizospheric relationships consist of the PGPRs that colonize the surface of the root, or superficial intercellular spaces of the host plant, often forming root nodules. The dominant species found in the rhizosphere is a microbe from the genus *Azospirillum*. Endophytic relationships involve the PGPRs residing and growing within the host plant in the apoplastic space.

Nitrogen Fixation

Nitrogen fixation is one of the most beneficial processes performed by rhizobacteria. Nitrogen is a vital nutrient to plants and gaseous nitrogen (N_2) is not available to them due to the high energy required to break the triple bonds between the two atoms. Rhizobacteria, through nitrogen fixation, are able to convert gaseous nitrogen (N_2) to ammonia (NH_3) making it an available nutrient to the host plant which can support and enhance plant growth. The host plant provides the bacteria with amino acids so they do not need to assimilate ammonia. The amino acids are then shuttled back to the plant with newly fixed nitrogen. Nitrogenase is an enzyme involved in nitrogen fixation and requires anaerobic conditions. Membranes within root nodules are able to provide these conditions. The rhizobacteria require oxygen to metabolize, so oxygen is provided by a hemoglobin protein called leghemoglobin which is produced within the nodules. Legumes are well-known nitrogen-fixing crops and have been used for centuries in crop rotation to maintain the health of the soil.

Symbiotic Relationships

The symbiotic relationship between rhizobacteria and their host plants is not without costs. For the plant to be able to benefit from the added available nutrients provided by the rhizobacteria, it needs to provide a place and the proper conditions for the rhizobacteria to live. Creating and maintaining root nodules for rhizobacteria can cost between 12–25% of the plant's total photosynthetic output. Legumes are often able to colonize early successional environments due to the unavailability of nutrients. Once colonized, though, the rhizobacteria make the soil surrounding the plant more nutrient rich, which in turn can lead to competition with other plants. The symbiotic relationship, in short, can lead to increased competition.

PGPRs increase the availability of nutrients through the solubilization of unavailable forms of nutrients and by the production of siderophores which aids in the facilitating of nutrient transport. Phosphorus, a limiting nutrient for plant growth, can be plentiful in soil, but is most commonly found in insoluble forms. Organic acids and phosphotases released by rhizobacteria found in plant rhizospheres facilitate the conversion of insoluble forms of phosphorus to plant-available forms such as $H_2PO_4^-$. PGPR bacteria include *Pseudomonas putida*, *Azospirillum fluorescens*, and *Azospirillum lipoferum* and notable nitrogen-fixing bacteria associated with legumes includes *Allorhizobium*, *Azorhizobium*, *Bradyrhizobium*, and *Rhizobium*.

Though microbial inoculants can be beneficial for crops, they are not widely used in industrial agriculture, as large-scale application techniques have yet to become economically viable. A notable exception is the use of rhizobial inoculants for legumes such as peas. Inoculation with PGPRs ensures efficient nitrogen fixation, and they have been employed in North American agriculture for over 100 years.

Pathogenic Roles

Studies conducted on sugar beet crops found that some root-colonizing bacteria were deleterious rhizobacteria (DRB). Sugar beet seeds inoculated with DRB had reduced germination rates, root lesions, reduced root elongation, root distortions, increased fungi infection, and decreased plant growth. In one trial the sugar beet yield was reduced by 48%.

Six strains of rhizobacteria have been identified as being DRB. The strains are in the genera *Enterobacter*, *Klebsiella*, *Citrobacter*, *Flavobacterium*, *Achromobacter*, and *Arthrobacter*. Due to a large number of taxonomic species yet to be described, complete characterization has not been possible as DRB are highly variable.

The presence of PGPRs has proven to reduce and inhibit the colonization of DRB on sugar beet roots. Plots inoculated with PGPRs and DRBs had an increase in production of 39% while plots only treated with DRBs had a reduction in production of 30%.

Biocontrol

Rhizobacteria are also able to control plant diseases that are caused by other bacteria and fungi. Disease is suppressed through induced systematic resistance and through the production of antifungal metabolites. *Pseudomonas* biocontrol strains have been genetically modified to improve plant growth and improve the disease resistance of agricultural crops. In agriculture, inoculant bacteria are often applied to the seed coat of seeds prior to being sown. Inoculated seeds are more likely to establish large enough rhizobacterial populations within the rhizosphere to produce notable beneficial effects on the crop.

Plant Growth-promoting Rhizobacteria

The rhizosphere supports the development and activity of a huge and diversified microbial community, including microorganisms capable to promote plant growth. Among the latter, plant growth-promoting rhizobacteria (PGPR) colonize roots of monocots and dicots, and enhance plant growth by direct and indirect mechanisms. Modification of root system architecture by PGPR implicates the production of phytohormones and other signals that lead, mostly, to enhanced lateral root branching and development of root hairs. PGPR also modify root functioning, improve plant nutrition and influence the physiology of the whole plant. Recent results provided first clues as to how PGPR signals could trigger these plant responses. Whether local and systemic, the plant molecular pathways involved remain often unknown. From an ecological point of view, it emerged that PGPR form coherent functional groups, whose rhizosphere ecology is influenced by a myriad of abiotic and biotic factors in natural and agricultural soils, and these factors can in turn modulate PGPR effects on roots. In this paper, we address novel knowledge and gaps on PGPR modes of action and signals, and highlight recent progress on the links between plant morphological and physiological effects induced by PGPR. We also show the importance of taking into account the size, diversity, and gene expression patterns of PGPR assemblages in the rhizosphere to better understand their impact on plant growth and functioning. Integrating mechanistic and ecological knowledge on PGPR populations in soil will be a prerequisite to develop novel management strategies for sustainable agriculture.

Photosynthetic terrestrial plants play key roles as ecosystem engineers. They contribute, for instance, to the establishment of specific microbial ecological niches in plant-based systems. This is particularly the case in the rhizosphere, i.e., the soil in contact with plant roots. Besides its role in plant anchorage in soil, absorption of water and ions, nutrient storage, and plant vegetative growth, the root system is in close contact with a wide range of soil microbial populations.

Despite their interactions with the biotic environment, the root system and its rhizosphere have received much less attention by plant physiologists than the rest of the plant. Plant roots exude a huge diversity of organic nutrients (organic acids, phytosiderophores, sugars, vitamins, amino acids, nucleosides, mucilage) and signals that attract microbial populations, especially those able to metabolize plant-exuded compounds and proliferate in this microbial habitat. Root exudates being the largest source of carbon supply within soil, the rhizosphere compartment houses a rich microbial community, comprising up to 1010 bacteria per gram of soil and encompassing a large diversity of taxa. The corresponding microbial community associated to plant roots can be referred to as the rhizo-microbiome. Its composition is distinct from that of the microbial community of the surrounding soil, a direct consequence of bacterial competition for nutrients liberated in the vicinity of plant roots. Since root exudate composition changes along the root system, according to stages of plant development and to plant genotypes, the rhizo-microbiome composition differs accordingly.

Within the rhizo-microbiome, some microorganisms can promote plant growth and provide better plant health through several indirect or direct mechanisms. Beneficial plant-microbe interactions are symbiotic interactions in which costs and benefits are shared by the plants and the microorganisms and can be categorized into two main types of interactions. First, mutualistic interactions correspond to intimate and mostly obligate interactions between microbes and a restricted range of compatible host plants. They generally lead to the formation of a structure specifically dedicated to the interaction (e.g., nodules during the symbiosis between nodulating rhizobia and Fabaceae, arbuscules in the endomycorrhizal symbiosis). Second, cooperations (also called associative symbioses) correspond to less obligate and specific interactions. They involve soil bacteria able to colonize the surface of the root system (and sometimes root inner tissues) and to stimulate the growth and health of the plant, and are referred to as plant growth-promoting rhizobacteria. Colonization of plant host roots by PGPR is heterogeneous along the root system; their competitiveness regarding this process is a sine qua non for plant growth promotion. In comparison to mutualistic symbionts, PGPR are thought to interact with a large range of host plant species and to encompass a huge taxonomic diversity, especially within the Firmicutes and Proteobacteria phyla. PGPR can enhance plant nutrition via associative nitrogen fixation, phosphate solubilization, or phytosiderophore production. They can improve root development and growth through the production of phytohormones or enzymatic activities, as well as favor the establishment of rhizobial or mycorrhizal symbioses. Others can protect the plant through inhibition of phytoparasites, based on antagonism or competition mechanisms, and by eliciting plant defenses such as induced systemic resistance. Some PGPR can also help plants withstand abiotic stresses including contamination by heavy metals or other pollutants; certain are even able to increase the capacity of plants to sequester heavy metals. Therefore, utilizing PGPR is a new and promising approach for improving the success of phytoremediation of contaminated soils.

Understanding and quantifying the impact of PGPR on roots and the whole plant remain challenging. One strategy is to inoculate roots with a PGPR in vitro and monitor the resulting effects on plant. This showed that many PGPR may reduce the growth rate of the primary root, increase the number and length of lateral roots, and stimulate root hair elongation in vitro. Consequently, the uptake of minerals and water, and thus the growth of the whole plant, can be increased. Some of these effects, including increased root and shoot biomass, are also documented for PGPR-inoculated plants growing in soil.

Impact of PGPR on Root System Architecture and Root Structure

Most terrestrial plants develop their root system to explore soil and find nutrients to sustain growth. Root is a complex organ made of distinct regions such as the root tip, root meristem, differentiation and elongation zones, and emerging lateral roots. These regions have distinct roles. For instance, root hairs are differentiated epidermal cells important for plant mineral nutrition, as inferred from gene expression studies and nutrient accumulation measurements. Root functional specificity is also reflected at the level of plant-microbe interactions. In Fabaceae for example, the root tip is the most important region to initiate the rhizobial colonization process leading eventually to the formation of a root nodule. In Poaceae, root hairs and lateral roots are preferentially colonized by PGPR, where they may express their plant beneficial properties. Root system architecture (RSA) integrates root system topology, the spatial distribution of primary and lateral roots, and the number and length of various types of roots. Several abiotic and biotic factors can influence RSA, including PGPR strains. PGPR modify RSA and the structure of root tissues mainly through their ability to interfere with the plant hormonal balance.

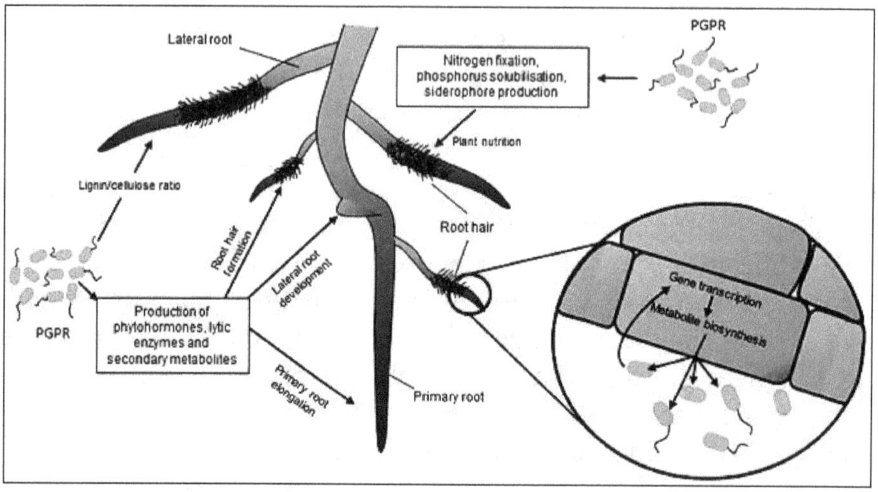

Impact of phytostimulating PGPR on RSA, nutrient acquisition and root functioning. PGPR can modulate root development and growth through the production of phytohormones, secondary metabolites and enzymes. The most commonly observed effects are a reduction of the growth rate of primary root, and an increase of the number and length of lateral roots and root hairs. PGPR also influence plant nutrition via nitrogen fixation, solubilization of phosphorus, or siderophore production, and modify root physiology by changing gene transcription and metabolite biosynthesis in plant cells.

PGPR Effects on RSA via Modulation of Host Hormonal Balance

Changes in RSA may result from interferences of PGPR with the main hormonal pathways involved in regulating plant root development: auxin, cytokinin, ethylene, and to a lesser extend gibberellin, and abscisic acid (ABA). The balance between auxin and cytokinin is a key regulator of plant organogenesis, and shapes root architecture. The auxin to cytokinin ratio can be affected by PGPR because they are able to produce a wide range of phytohormones, including auxins and cytokinins, and secondary metabolites that can interfere with these hormonal pathways.

Many PGPR are able to produce phytohormones and secondary metabolites interfering with the plant auxin pathway, such as auxins, 2,4-diacetylphloroglucinol (DAPG), and nitric oxide (NO). Indole-3-acetic acid (IAA) is the best-characterized auxin produced by many plant-associated bacteria, including PGPR. Exogenous IAA controls a wide variety of processes in plant development and plant growth: low concentrations of IAA can stimulate primary root elongation, whereas high IAA levels stimulate the formation of lateral roots, decrease primary root length and increase root hair formation. IAA is usually synthesized by rhizobacteria from tryptophan, which is found at different concentrations in root exudates according to plant genotype. In PGPR strains, several IAA biosynthetic pathways have been described depending on the metabolic intermediates. The indole-3-pyruvate decarboxylase (encoded by the ipdC/ppdC bacterial gene) is a key enzyme involved in the indolepyruvic acid pathway. Effects of ipdC mutants on plant root morphology are often altered in comparison to those of wild-type strains.

Plant growth promotion by PGPR can also result from indirect stimulation of the plant auxin pathway. For example, several PGPR strains like Azospirillum brasilense have a nitrite reductase activity and consequently are able to produce NO during root colonization. NO is involved in the auxin signaling pathway controlling lateral root formation. DAPG is a well-known antimicrobial compound produced by biocontrol fluorescent pseudomonads. At lower concentrations, DAPG can also be a signal molecule for plants, inducing systemic resistance, stimulating root exudation, and enhancing root branching. DAPG can interfere with an auxin-dependent signaling pathway and thus modify RSA. Indeed, applications of exogenous DAPG, at a concentration around 10 μM, inhibited primary root growth and stimulated lateral root production in tomato seedlings. Furthermore, roots of an auxin-resistant diageotropica mutant of tomato displayed reduced DAPG sensitivity.

The growth-promotion effect of auxin or auxin-like compounds by PGPR may require functional signaling pathways in the host plant. To test that hypothesis, one could use a host plant defective at a particular step of the hormone-signaling pathway and assess whether PGPR inoculation complements or not the effect of the mutation. This strategy requires the use of model plant such as Arabidopsis, the only biological system that provides to date enough documented mutant plants. Consistent with that, some Arabidopsis auxin-signaling mutants failed to show the typical root architecture changes in response to the beneficial rhizobacterium Phyllobacterium brassicacearum STM196. However, auxin content was not increased in roots upon inoculation with Phyllobacterium brassicacearum STM196, ruling out the potential implication of auxin of bacterial origin. Nevertheless, the use of Arabidopsis DR5::GUS reporter line, whose expression is restricted to the root meristem where the auxin maximum is located, showed a change of expression pattern in response to STM196 inoculation. GUS staining appeared more intense on a wider region of the root tip as well as in the vasculature, suggesting that there was a change of auxin distribution in the root in response to STM196 inoculation, even though this strain is a low auxin producer. Interestingly, a similar observation was made when Arabidopsis plantlets were inoculated with the PGPR Bacillus subtilis

GB03, which emits volatile organic compounds (VOCs), or with Pseudomonas fluorescens WCS417.

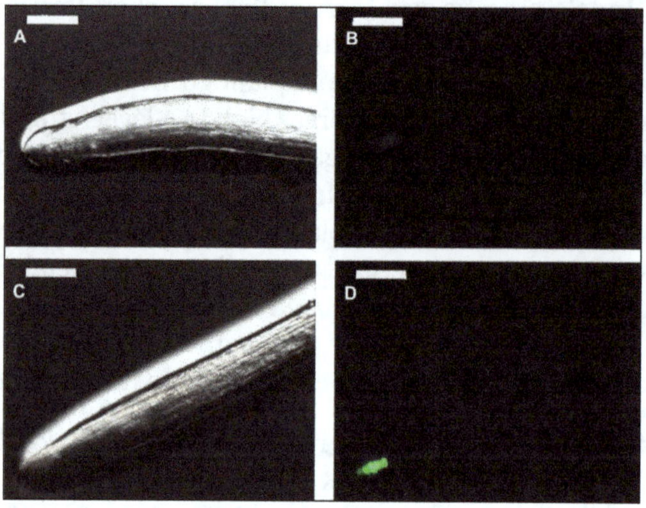

PGPR-mediated changes in RSA may relate to modifications of auxin content in roots. Six-day-old Arabidopsis plantlets expressing the GFP gene under the control of the auxin-sensitive DR5 artificial promoter were inoculated (C, D) or not (A, B) with the PGPR Phyllobacterium brassicacearum STM196. Six days later, root tips were observed under normal light (A, C) or UV light (B, D) with a microscope (Z16APO, Leica). Scale bars represent 200 µm. Inoculation by STM196 modified root traits such as root hair elongation and primary root growth, which coincided with an increase in GFP signal in the root tip in inoculated (D) compared with control plants (B). These observations confirm previous results with a different Arabidopsis DR5 reporter line.

Cytokinin production (especially zeatin) has been documented in various PGPR like Arthrobacter giacomelloi, Azospirillum brasilense, Bradyrhizobium japonicum, Bacillus licheniformis, Pseudomonas fluorescens, and Paenibacillus polymyxa. Cytokinins stimulate plant cell division, control root meristem differentiation, induce proliferation of root hairs, but inhibit lateral root formation and primary root elongation. Inoculation of plants with bacteria producing cytokinin has been shown to stimulate shoot growth and reduce the root to shoot ratio. Bacterial genes involved in cytokinin biosynthetic pathways have been identified in silico but their role has not yet been validated through functional analyses. Consequently, the contribution of cytokinin production by PGPR to RSA modifications remains speculative.

Ethylene is another key phytohormone, which inhibits root elongation and auxin transport, promotes senescence and abscission of various organs, and leads to fruit ripening. Ethylene is also involved in plant defense pathways. This phytohormone can be produced in small amounts from the precursor methionine by some PGPR, like Azospirillum brasilense. The ability of Azospirillum brasilense to produce ethylene presumably promotes root hair development in tomato plants. Indeed, exogenous ethylene supply to the plant mimicked the effect of Azospirillum brasilense inoculation, while the addition of an ethylene biosynthesis inhibitor blocked this effect. Actually, PGPR are more widely able to lower plant ethylene levels through deamination of 1-aminocyclopropane-1-carboxylic acid (ACC). Many genomes of PGPR do contain a gene (acdS) coding for an ACC deaminase, which degrades ACC into ammonium and α-ketobutyrate. By lowering the abundance of the ethylene precursor ACC, the PGPR AcdS activity is thought to decrease root ethylene production, which can in turn alleviate the repressing effect of ethylene on root growth. Despite being widely accepted and supported by experimental data, the model raises issues that have not been well addressed yet. The first one deals with ethylene production within roots. Light is promoting ethylene biosynthesis, providing there is a sufficient CO_2 supply for shoots. Exposition of roots to light was shown to trigger an increase in ethylene production.

In soil however, roots are sheltered from light, suggesting that this organ might not be able to synthesize large amounts of ethylene. In agreement with that, Fabaceae roots did produce ethylene in response to rhizobial colonization in presence of light, but less when they were in the dark. Secondly, there is a regulation of ethylene synthesis by a feedback loop. This loop should stimulate ethylene biosynthesis when the level of ACC is low. Unless PGPR disconnect that feedback loop, lowering ACC content would eventually result in stimulation of ethylene production. There is no indication yet how the feedback loop would work in presence of a PGPR. Last but not least, if ethylene plays a key role during the plant-PGPR interaction, one would expect that either plant ethylene mutants or impaired AcdS activity in the bacteria would result in clear disturbance of the plant responses to bacteria. However, minor effects on RSA were observed when plants were inoculated with an acdS bacterial mutant, or when plants affected in their ethylene signaling pathway were inoculated with wild-type PGPR. It suggests that ethylene participates to the root architecture response but is not a key player. Taken together, the functional importance of the bacterial ACC deaminase function needs further clarification. One hypothesis could be that AcdS contributes to the fine-tuning of ethylene biosynthesis during the plant-PGPR cooperation.

Several reports have revealed that PGPR are able to produce ABA or gibberellic acid, or to control the level of these hormones in plants. The first one, ABA, is well known for its involvement in drought stress. During water stress, increase in ABA levels causes closing of stomata, thereby limiting water loss. However, this hormone also plays different roles during lateral root development. Inoculation with Azospirillum brasilense Sp245 led to an increase of ABA content in Arabidopsis, especially when grown under osmotic stress. Gibberellins promote primary root elongation and lateral root extension. Production of gibberellins has been documented in several PGPR belonging to Achromobacter xylosoxidans, Acinetobacter calcoaceticus, Azospirillum spp., Azotobacter spp., Bacillus spp., Herbaspirillum seropedicae, Gluconobacter diazotrophicus and rhizobia. Application of gibberellic acid on maize, at a concentration similar to that produced by Azospirillum, promotes root growth; furthermore, gibberellin content increases in maize root inoculated with Azospirillum.

In addition to playing a role in plant RSA, these two hormones are involved in plant defense mechanisms. Thus, PGPR producing these hormones may modulate the hormonal balance involved in plant defense, including the jasmonate and salicylic acid pathways.

Although the production of hormones by PGPR has been well described, the genetic determinants involved in their biosynthesis remain largely unknown and bacterial mutants affected in hormone biosynthesis are mostly lacking. Consequently, the involvement of hormones of bacterial origin in the modulation of plant hormonal balance has not been fully demonstrated.

Modification of Root Cell Wall and Root Tissue Structural Properties by PGPR

Many PGPR can lead to modifications of the chemical composition and therefore structural properties of root cell walls. For example, the biocontrol agent Bacillus pumilus INR-7 is able to enhance lignin deposition in pearl millet epidermal tissues, and this plant defense response appears much more rapidly in PGPR-primed plants infected by the pathogen Sclerospora graminicola compared to non-primed plants. The sole inoculation of INR-7 led to callose apposition. Although the precise location of these deposited polymers was not investigated, it is possible that their enhanced accumulation may participate to pathogen inhibition and disease suppression. A similar response was also triggered by Bacillus pumilus SE34 and Bacillus subtilis UMAF6639 when inoculated to respectively pea and melon roots. In both cases, it led to enhanced fungal pathogen tolerance. Inoculation of Pseudomonas fluorescens 63-28R to pea roots induced accumulation of lignin in root cells and inhibited colonization by the oomycete Pythium ultimum. The same result was observed with a Pseudomonas putida strain inoculated on bean roots. These cell wall modifications have been reported in the case of PGPR that protect plants against phytopathogens by activating ISR plant defense responses. One of the consequences of ISR is thus the reinforcement of the cell wall through enhanced lignin synthesis and callose apposition, which restricts the progression of phytopathogens through plant tissues.

Modifications of the chemical composition of root cell walls are also triggered by PGPR that directly promote plant growth. Through the analysis of the infrared spectral characteristics of crude cell wall preparations of maize roots, El Zemrany et al. concluded that roots inoculated with Azospirillum lipoferum CRT1 had lower lignin content than uninoculated ones. This result contrasts with those aforementioned for biocontrol agents. Nevertheless, lower lignin content may facilitate cell elongation, and therefore overall root elongation. Similarly, Azospirillum irakense produces pectate lyases that are capable of degrading the pectate content of root cell wall and might allow its progression between root cortex cells and its functioning as an endophyte. Up to now, the impact on plant lignin content of PGPR that are both inducing ISR and promoting root growth has not been clarified.

Modifications of root cell wall ultrastructure are thought to result mainly from PGPR-triggered changes in plant gene expression. Indeed, Bacillus subtilis GB03 promotes Arabidopsis growth by producing VOCs that were shown to modulate the expression of 38 genes with known functions associated with cell wall structure. Among them, 30 were implicated in cell wall expansion or loosening. The endophytic PGPR Azospirillum irakense was also shown to stimulate the expression of polygalacturonase genes in inoculated rice roots.

Chemical mediators involved in the effects of PGPR on root cell walls have received little attention. A single report has indicated that the exogenous addition of auxins enhanced the induced polygalacturonase activities observed in Azospirillum irakense inoculated rice roots.

Systemic Effects of PGPR on Whole Plant Physiology and Functioning

In addition to their effects on root tissues, PGPR can modify the physiology and functioning of plant tissues located at a substantial distance from the colonized sites, such as shoots. Two types of mechanisms are involved. On the one hand, some PGPR can enhance nutrient availability/uptake for plant roots. Stimulation of plant nutrition will lead to modifications in primary metabolism and consequently will contribute to enhance growth. On the other hand, certain PGPR trigger specific systemic responses, mostly by unknown signaling mechanisms. High-throughput analyses of plant transcriptomic and metabolomic responses have evidenced the effects of PGPR on plant gene expression and metabolite accumulation, respectively. These results highlight the extensive effect of PGPR on whole plant physiology and functioning, and provide clues to understand the systemic effect of PGPR.

Impact of PGPR on Plant Nutrition

The impact of PGPR on plant nutrition may result from effects on plant nutrient uptake and on plant growth rate. It is indeed commonly hypothesized that nutrient uptake is increased as a consequence of increased root surface area triggered by PGPR. However, root ion transporters are under the control of regulatory processes that adjust their activity to the plant nutritional demand, so that regulations of root development and ion transporter activities are antagonistically coordinated to maintain steady nutrient acquisition rate. Hence, PGPR must interfere with pathways coordinating plant development and plant nutrition to elicit both increased nutrient acquisition rate and plant growth promotion.

Plant growth-promoting rhizobacteria can directly increase nutrient supply in the rhizosphere and stimulate ion transport systems in root. With regards to increased nutrient supply, two main types of bacterial activities can be considered. Firstly, phosphate solubilization is one key effect of PGPR on plant nutrition. Soils generally contain a large amount of phosphorus, which accumulates in the wake of regular fertilizer applications, but only a small proportion of the latter is available for plants. Plants are able to absorb on their own mono and dibasic phosphate; organic or insoluble forms of phosphate need to be mineralized or solubilized by microorganisms, respectively. Many PGPR – such as Pseudomonas, Bacillus, Rhizobium – are able to dissolve insoluble forms of phosphate. Two main processes exist: acidification of the external medium through the release of low molecular weight organic acids (such as gluconic acid) that chelate the cations bound to phosphate, and production of phosphatases/phytases that hydrolyse organic forms of phosphate compounds. Secondly, many associated bacteria can fix N_2 so that they could provide nitrogen to the plant. Evidence in favor of the participation of PGPR to the plant N budget has been reported for several plants, especially sugar cane. However, the impact of N_2-fixation by PGPR is still debated and is rarely credited for the stimulation of plant growth. In addition, non-fixing rhizobacteria can promote plant growth, showing that N provision is not obligatory for plant growth promotion. For instance, Phyllobacterium brassicacearum STM196 is unlikely to fix N_2 while it promotes the growth of canola and Arabidopsis.

With regards to the impact of PGPR on nutrient uptake systems, only very few studies have been published so far. Inoculation of canola with Achromobacter sp. strain U80417 resulted in an increase of both NO_3^- and K^+ net influx rates per root surface area unit. The net H^+ efflux was also enhanced, so that increased NO_3^- and K^+ uptake rates may be part of a general mechanism leading to increased ion uptake rate, similar to energization of nutrient transport by enhanced proton pump activity. In favor of this hypothesis, acidification of the rhizosphere has also been reported with Arabidopsis exposed to the VOC-emitting Bacillus subtilis strain GB03.

In Arabidopsis, NO_3^- uptake measurement in response to PGPR, over time, can lead to contradictory results: NO_3^- influx was increased in seedlings, upon 24 h-inoculation with Phyllobacterium brassicacearum STM196, while it was reduced 7 days later. However, it is hard to draw a firm conclusion as the net NO^{-3} uptake rate remained unknown since ion efflux was not measured in these experiments. Except for the NRT2.5 and NRT2.6 genes, the accumulation of transcripts of nitrate and ammonium transporters was very slightly or not significantly changed upon Phyllobacterium brassicacearum STM196 inoculation. Interestingly, these two genes have recently been shown to be required in Arabidopsis growth promotion by this PGPR. Since these two genes code for two plasma membrane-localized NO_3^- transporters, this discovery raises the question of the interactions between N nutrition and plant development in PGPR-inoculated plants. The NRT2.5 and NRT2.6 genes are predominantly expressed in shoots. Their role in Phyllobacterium brassicacearum STM196 plant growth promotion and root architecture modification are not linked to changes in NO_3^- uptake rate or NO_3^- distribution between roots and shoots, suggesting an involvement in N-signaling rather than a direct role in N-metabolism.

Evidence in favor of a regulation of ion transporters at a transcriptional level by PGPR has been obtained in studies with Bacillus subtilis GB03. This strain induces concomitant down- and up-regulation of HKT1 expression in roots and shoots of Arabidopsis seedlings, respectively. In the shoots, HKT1 functions in phloem tissues to retrieve Na^+ from the xylem and in the roots it is involved in Na^+ uptake. The differential regulation of HKT1 expression in roots and shoots resulted in reduced accumulation of Na^+ and increased accumulation of K^+ in both organs of GB03-inoculated seedlings under salt-stress conditions. Consistent with the effect of GB03 on HKT1, GB03 failed to rescue salt-stressed hkt1 mutant seedlings from elevated Na^+ accumulation.

Volatile organic compounds emitted by GB03 also activate the plant's iron acquisition machinery leading to increased iron assimilation. Firstly, this PGPR leads to acidification of the rhizosphere, both directly due to chemical effects of some unidentified VOCs and indirectly through increased root proton efflux. Secondly, GB03 up-regulates the expression levels of FRO2 and IRT1 genes, coding respectively for a Fe_3^+ chelate reductase and a Fe_2^+ transporter. As a result, GB03-exposed Arabidopsis has enhanced ferric chelate reductase activity and increased iron content. Finally, it has been shown that this PGPR induces the expression of the FIT1 transcription factor that regulates positively FRO2 and IRT1 expressions. The fact that GB03 fails to increase root ferric reductase activity and plant iron content in Arabidopsis fit1 mutants shows that PGPR can modify indirectly ion uptake by interfering with plant regulatory processes that control ion transporter expressions and activities.

Impact of PGPR on Plant Transcriptome

Targeted or genome-wide analyses of plant gene expression following root inoculation by PGPR were reported with various bacterial models: phytostimulating PGPR, endophytes and PGPR exerting a biocontrol activity. Inoculation of the phytostimulator Pseudomonas putida MTCC5279 triggered overexpression of 520 genes and repression of 364 genes (threefold changes) in leaves of Arabidopsis; upregulated genes were involved in maintenance of genome integrity, growth hormone and amino acid syntheses, ABA signaling and ethylene suppression, Ca_2^+ dependent signaling and induction of ISR. On rice, a recent study performed with Azospirillum points towards association specificity. The targeted expression of ethylene receptors was followed after inoculation of Azospirillum brasilense Sp245 on two rice cultivars of contrasted ability to gain nitrogen from biological nitrogen fixation. Seedlings of cultivar IR42, which enabled higher nitrogen fixation, also displayed higher expression of ethylene receptors compared to cultivar IAC 4440. The transcript accumulation of all ethylene receptors might be necessary for the establishment of a beneficial association between the plant and the bacteria.

As for endophytes, differential colonization of rice roots was observed with an Azoarcus PGPR. In a less compatible interaction, a slight defense response occurred and was accompanied by the induction of pathogenesis-related proteins and proteins sharing domains with receptor-like kinases induced by pathogens; those proteins were also induced by a jasmonate treatment. Inoculation of rice roots with the endophytic PGPR

Herbaspirillum seropedicae triggered the expression of genes responsive to auxin and ethylene and the repression of the defense-related proteins PBZ1 and thionins. These studies suggest that endophytes modulate plant defense responses during colonization.

Plants treated with biocontrol PGPR, usually belonging to the Pseudomonas genus, are more resistant to subsequent infections by bacterial or fungal pathogens. In Arabidopsis, this rhizobacteria-mediated ISR requires sensitivity to jasmonate and ethylene, and the regulators MYC2, NPR1, and MYB72 played a central role in this signaling. One of the earliest transcriptomic study performed with Pseudomonas fluorescens WCS417r indicated that bacteria elicited a substantial change in the expression of 97 genes in roots whereas none of the approximately 8,000 genes tested showed a consistent change in expression in the leaves. Subsequent studies on Arabidopsis reported an increase of defense-related transcripts, including PR-related proteins, in shoots of bacterized plants compared to untreated shoots. Interestingly, the ISR induced by Pseudomonas fluorescens SS101 was recently reported to be dependent on salicylic acid signaling and not on jasmonic acid and ethylene signaling; moreover, a prominent role of camalexin and glucosinolates in the ISR was proposed. In wheat, bacterization with Pseudomonas fluorescens Q8r1-96 also triggered the accumulation of defense-related transcripts and neither DAPG nor the type three secretion systems were key single factors in the expression of these genes. The establishment of beneficial associations requires mutual recognition and substantial coordination of plant and microbial responses and consequently beneficial microbes modulate plant immunity.

Impact of PGPR on Plant Metabolome

Several studies have addressed the metabolomic changes triggered by PGPR inoculation, by analyzing metabolite contents of root exudates, root tissues and shoots under normal or stressful conditions. Some studies have shown that PGPR can elicit changes in the activities of root enzymes involved in the production of metabolites, especially flavonoids, leading to changes in the pattern of root exudation. Some Azospirillum PGPR stimulated by up to one-third the level of carbon compounds exuded from roots. Moreover, compounds of microbial origin, such as phenazines and DAPG, could enhance total net efflux of amino acids in plant species. On soybean roots, the PGPR Chryseobacterium balustinum Aur9 influences flavonoids exudation. PGPR strains from Chryseobacterium or Azospirillum may influence flavonoid exudation by Fabacea roots. This property can be important for the design of mixed inoculants that will include a PGPR strain promoting flavonoid exudation together with rhizobia that will respond to plant flavonoids.

In addition to effects on root exudates, PGPR can trigger modifications of metabolite composition of the whole plant. For instance, rice plants inoculated with Herbaspirillum seropedicae showed higher shoot contents in malate and in key amino acids than those of control plants. Many more studies focused on modifications of secondary metabolites. Elicitation of isoflavone accumulation was observed on soybean inoculated with various PGPR, either by increasing the total isoflavone content in seedlings or by causing an asymmetric distribution of isoflavones throughout the plant. Increase in the content of several alkaloid and terpenoid compounds of pharmaceutical relevance was demonstrated in medicinal plants following PGPR inoculation. Recent studies investigated the early impact of several Azospirillum strains on root and shoot secondary metabolite profiles of maize and rice; analysis of secondary metabolites of two maize cultivars, inoculated by three different Azospirillum strains under greenhouse conditions, revealed major qualitative and quantitative modifications of the contents of secondary metabolites, especially benzoxazinoids. In the same way, a metabolic profiling approach of two rice cultivars inoculated with two different Azospirillum strains under gnotobiotic conditions, showed that profiles of secondary metabolites were modified with phenolic compounds such as flavonoids and hydroxycinnamic derivatives being the main metabolites affected. Both studies evidenced a specific response, as plant metabolic changes differed according to the Azospirillum strain-cultivar combination. Moreover, PGPR applied to the roots can affect the composition of secondary metabolites in shoots, pointing towards systemic effects.

Accumulation of secondary compounds was also modified in several plants inoculated with consortia containing arbuscular mycorrhizal fungus and PGPR. Blumenin accumulation triggered by Rhizophagus irregularis (formerly Glomus intraradices) in barley and wheat roots was increased when a rhizosphere bacterium was applied with the fungus. Leaf secondary metabolites (total phenols and ortho dihydroxy phenols), as well as leaf mineral content (phosphorus, potassium, zinc, copper, and iron) were maximal when Begonia malabarica or Solanum viarum were inoculated with consortia containing two fungi and a Bacillus coagulans strain. Field-inoculation of maize with selected strains of Pseudomonas, Azospirillum or Rhizophagus/Glomus, or with these strains combined two by two or all three together, led to qualitative and quantitative modifications of root secondary metabolites, particularly benzoxazinoids and diethylphthalate. These modifications depended on fertilization level and on the type of microorganisms inoculated. The three selected strains gave distinct results when used alone, but unexpectedly all microbial consortia gave somewhat similar metabolic responses.

Plant growth-promoting rhizobacteria can help plants to withstand saline stress, a feature that may be linked to accumulation of specific metabolites. A higher level of proline content was reported in inoculated Bacopa monnieri, as well as higher accumulation of glycine betaine-like quaternary compounds in rice inoculated with Pseudomonas pseudoalcaligenes. Similarly, Arabidopsis inoculation with the VOC-emitting strain Bacillus subtilis GB03 induced strong plant accumulation of glycine betaine and its precursor choline, and GB03-induced drought tolerance was lost in the xipotl mutant of Arabidopsis with reduced choline production. Alleviation of cold stress was demonstrated for Burkholderia phytofirmans PsJN on grapevine; this endophytic strain promotes plant post-chilling recovery by improving acclimation to cold. This is accompanied by accumulation of stress-related metabolites such as proline, malondialdehyde and aldehydes (known as lipid peroxidation markers), hydrogen peroxide, and by higher expression of defense- and cold-related genes. Bacterization resulted in a 1.2-fold increase in starch content and in a two-fold increase in total soluble sugars, with sugars known to be involved in low-temperature tolerance (glucose, sucrose, and raffinose) displaying higher concentrations in treated plantlets. Independently of temperature, inoculation also enhanced phenolic content.

Expression of Plant-beneficial Functions of PGPR in the Rhizosphere

One PGPR strain can harbor several plant-beneficial properties, which may be co-regulated or not. Within the rhizosphere, the expression of PGPR's plant-beneficial properties is affected by both abiotic factors (like pH, oxygen, clay mineralogy, heavy metals, etc.) and biotic factors (i.e., compounds produced by plants or the rhizo-microbiome) that can lead to distinct expression patterns in space and time, possibly with different effects on host plant.

Regulation of PGPR Functions by Root Exudates

Through the release of root exudates, plants can impact bacterial gene expression, especially genes encoding plant-beneficial traits. Composition of root exudates is dependent upon intra and inter-specific genetic variability, plant developmental stage and soil abiotic factors. One of the major studies aiming at analyzing the impact of root exudates variability on bacterial gene expression was carried out on phlA, involved in DAPG biosynthesis, in Pseudomonas protegens (formely Pseudomonas fluorescens) CHA0. The expression of phlA was increased fourfold in the rhizosphere of monocots (maize and wheat) compared to the rhizosphere of dicots (bean and cucumber). The analysis of six maize cultivars also revealed that phl expression and hence biocontrol activity could be affected by plant. Specific components of root exudates, notably sugars, were shown to affect the production of antimicrobial compounds, such as DAPG, pyoluteorin and pyrrolnitrin by fluorescent pseudomonads, with some strain-dependent effects. Among 63 plant compounds related to defense or development, or involved in plant-microbe interactions (flavonoids, phenolic acids, phytohormones, etc.), many could modulate the expression of phlA and pltA in Pseudomonas protegens CHA0. No specific chemical structures were identified that generally induced or repressed phlA or pltA expression. Umbelliferone led to the strongest inhibition of phlA; salicylate, jasmonate, and methyl jasmonate, all slightly reduced phlA expression, whereas the plant hormone IAA induced phlA expression. None of these compounds had an effect on pltA expression whereas a previous study reported repression of both DAPG and pyoluteorin biosynthesis genes by salicylate.

1-Aminocyclopropane-1-carboxylic acid deamination (encoded by acdS) is another bacterial function that may be differentially expressed according to plant genotypes. Indeed, in vitro experiments demonstrated that some compounds present in root exudates tightly control acdS expression. First, ACC, the precursor of ethylene that is metabolized by AcdS, can positively regulate acdS expression. Second, leucine, by inhibiting oligomerization of the Lrp-type regulator AcdR, prevents transcription of acdS leading to a decrease of ACC deaminase activity in Pseudomonas putida UW4 and in Azospirillum lipoferum 4B. Finally, carbon sources can also influece acdS transcription.

Bacterial IAA biosynthesis mostly depends on tryptophan-related pathways. The main source of tryptophan for PGPR is root exudates. Measurement of tryptophan bioavailability from graminaceous roots (Avena barbata) indicated that this amino acid is abundant at the emergence of secondary roots. In the absence of exogenous tryptophan supply, bacterial IAA biosynthesis is insignificant. Next to being an IAA precursor, tryptophan also plays an important role in regulating positively the ipdC/ppdC gene. Other root-exuded amino acids like tyrosine and phenylalanine can also induce ipdC/ppdC expression. Besides amino acids, plant roots release compounds like vitamins (e.g., pyridoxine and nicotinic acid) and organic acids (e.g., phenylacetic acid and prephenic acid). All these compounds increase significantly IAA production in Azospirillum brasilense Sp245.

Metabolites present in root exudates can thus specifically modulate the expression of key genes involved in plant-beneficial functions. Consequently, specific physiological responses of the plant are dependent on the PGPR strain/plant cultivar combination.

Regulation of PGPR Functions by Microbial Signals

Plant growth-promoting rhizobacteria exchange several types of cell-to-cell communication signals between each other and with other rhizosphere-inhabiting bacteria and fungi, i.e., quorum-sensing (QS) signals that allow bacteria to monitor their density and to coordinate gene expression only when a quorum of cells is achieved and other bacterial signals that regulate gene expression independently of the cell density.

Quorum-sensing relies on the synthesis and perception of small diffusible molecules, such as N-acyl-homoserine lactones (AHLs). In fluorescent pseudomonads, colonization properties and biosynthesis of antimicrobial metabolites, such as phenazines, is often subjected to an AHL-based QS regulation. Production of pyrrolnitrin in Serratia plymuthica HRO-C48, a strain isolated from the rhizosphere of oilseed rape and able to protect crops against Verticillium wilt, is also under QS regulation. In S. plymuthica G3, an endophytic strain, QS positively regulates antifungal activity, production of exoenzymes, but negatively regulates IAA production. Among the genus Azospirillum, only a few strains belonging to the lipoferum species and isolated from rice, display the ability to produce AHL signals. In the rice endophyte Azospirillum lipoferum B518, AHL inactivation abolishes pectinase activity, increases siderophore synthesis and reduces IAA production (in stationary phase) but no effect is observed on cellulase activity and on the phytostimulatory effect. Moreover, a proteomic approach indicates that AHL-based QS regulation in Azospirillum is rather dedicated to control functions linked to rhizosphere competence and adaptation to plant roots.

Interestingly, several studies have shown that bacterial communication of a specific bacterial population could be jammed by other microbes; indeed, some soil bacteria can inactivate AHL (notably members of the genus Bacillus), whereas others can intercept AHL or can act as a physical barrier preventing their diffusion. Consequently, other members of the bacterial rhizosphere community can compromise expression of biocontrol traits in PGPR. Conversely, cross-talk between species using the same AHL signal or a structurally-related AHL can occur in natural habitats and was evidenced in the rhizosphere of wheat and tomato. Finally, plant compounds designated as AHL-mimics can also interfere with bacterial QS and may influence the expression of plant-beneficial functions. Some Pseudomonas fluorescens strains unable to synthetize AHLs but possessing the cognate receptor may even recognize a plant compound to trigger expression of genes involved in biocontrol properties.

Exometabolites produced by microbial populations including pathogenic fungal strains can also affect PGPR plant-beneficial properties. For instance, fusaric acid produced by Fusarium oxysporum represses the production of DAPG in the biocontrol strain Pseudomonas protegens CHA0. Next to their antifungal effect, some Pseudomonas-produced compounds can influence gene expression of biocontrol traits in pseudomonads. Indeed, in Pseudomonas protegens strains CHA0 and Pf-5, DAPG and pyoluteorin productions are influenced by positive autoregulation; moreover, DAPG and pyoluteorin mutually inhibit one another's production. In order to determine if DAPG could act as a signal on other PGPR strains than those of the fluorescent Pseudomonas group, a differential fluorescence induction promoter-trapping approach based on flow cytometry was developed on Azospirillum. Using this approach DAPG was shown to enhance expression of a wide range of Azospirillum brasilense genes, including genes involved in phytostimulation. Four of them (i.e., ppdC, flgE, nirK, and nifX-nifB) were upregulated on roots in the presence of Pseudomonas fluorescens F113 compared with its DAPG-negative mutant. Accordingly, Pseudomonas fluorescens F113 but not its DAPG-negative mutant enhanced the phytostimulatory effect of Azospirillum brasilense Sp245 on wheat. Thus, DAPG can act as a signal by which some beneficial pseudomonads may stimulate plant-beneficial activities of Azospirillum PGPR. This finding is also relevant in the context of inoculation with microbial consortia, in which different types of PGPR may be combined. The number of studies investigating the efficacy of such combined inoculations is growing, with variations in the number of microorganisms and the nature of the combinations (PGPR strains only, PGPR and fungi, etc).

Field inoculation of sorghum with fluorescent Pseudomonas strains alone or in combination with arbuscular mycorrhizal fungi showed a better effect when in presence of the latter. Field inoculation of maize with a consortium consisting of two PGPR (Azospirillum lipoferum CRT1 and Pseudomonas fluorescens F113) and one mycorrhizal strain (Rhizophagus irregularis/Glomus intraradices JJ291) showed an increase of root surface, root volume and number of roots, although data were not statistically significant compared to the single Rhizophagus inoculation. Modification of one member of this consortium (three different Azospirillum strains were tested) could lead to significant modification of maize growth. Further studies are needed to describe the synergistic effects between beneficial microorganisms at a molecular scale and to analyse the expression of plant-beneficial functions when consortia are used.

Ecology of PGPR Populations and Impact on Root System Functioning

Many studies have deciphered the mechanisms of action of PGPR using one individual strain and one host plant. But in reality, as described above, PGPR strains are not acting individually in the rhizosphere but rather as part of bacterial communities, in which cell communication signals may coordinate the activities of all individual strains. Indeed, a vast array of PGPR populations displaying co-occurring plant-beneficial activities and that may share between each other antagonistic or synergistic effects are interacting with a same host plant. When analysing plant growth-promoting effects, it is thus important to integrate the complexity of the interactions between PGPR populations within the rhizo-microbiome. To do so, functional ecology approaches are needed, in which the relations between the size, diversity and activities of PGPR assemblages in the rhizosphere are taken into account. This is of particular importance when assessing the effect of various environmental factors, including that of plant genotype.

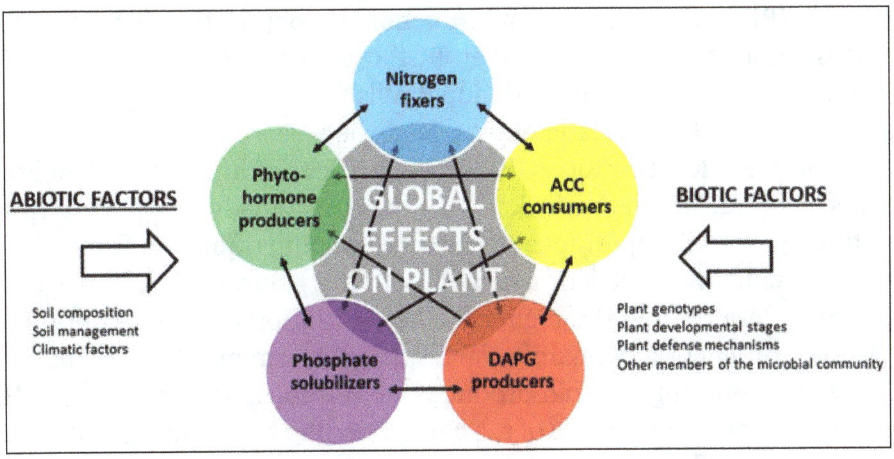

Implementation of plant-growth promoting traits in PGPR functional groups. Selected PGPR functional groups are represented by different colored circles. The resulting effect of all PGPR functional groups on the plant is symbolized by the gray circle. Abiotic and biotic factors may influence the activity of each functional group. Solid arrows represent potential interactions (inhibition, signaling, etc.) between members of the functional groups, which may impact on the size, diversity and activity of these groups.

PGPR Ecology in the Rhizosphere: From Individual Strains to Functional Groups

Plant growth-promoting rhizobacteria strains occur in various taxonomic groups, and these different taxonomic groups may be present simultaneously in a given soil. This suggests that taxonomically-contrasted PGPR strains may coexist in soil and colonize a same rhizosphere, along with all non-PGPR members of the bacterial community. This possibility has been documented repeatedly, especially when characterizing the taxonomic status of bacterial isolates selected based on their positive effect on plant growth or health, their ability to inhibit phytopathogens, or the occurrence of a particular gene or property of relevance for PGPR effect. In fact, this possibility seems to be the rule rather than the exception. PGPR populations contributing to a same type of function (i.e., ISR, nitrogen fixation, nutrient solubilization, plant development enhancement, etc.) belong to a same functional group. Functional group approaches can be implemented when specific genes are documented. For instance, nitrogen fixers can be assessed using the nifH gene, which encodes the dinitrogenase subunit of the nitrogenase. Its sequence is well conserved within the functional group and it is commonly used as marker to monitor the size and diversity of the diazotrophic community. Some of these PGPR functional groups are taxonomically narrow, such as the Pseudomonas DAPG producers. In contrast, others are much more diversified, and certain bacterial functional groups may also comprise both PGPR and non-PGPR strains. For instance, nitrogen fixers include PGPR as well as mutualistic symbionts and even a few pathogens.

When considering PGPR-plant relationship in fields, the co-occurrence of genetically contrasted PGPR strains from a same functional group in the rhizosphere has two consequences. First, the activity of a given PGPR functional group corresponds to the resulting contributions of all active individual cells from each type of bacterium within the functional group. If synergistic effects occur between the PGPR populations, the expected performance level for the PGPR function might be higher than if only one type of strain was involved. On this basis, knowing the size of the functional group will help understand the potential importance of the corresponding function. Indeed, for functions leading to enhanced nutrient availability to the plant, such as nitrogen fixation or phosphorus solubilization, the higher the better. For others where optimality matters, such as the production of auxinic signals, the performance level of the functional group will need fine-tuning to avoid production levels too small or too great. How this is ecologically regulated at the scale of the corresponding functional group is unknown, but it raises the possibility of co-evolutionary patterns. To bridge the gap between the potential of a plant-beneficial PGPR function and its actual implementation by PGPR strains, the regulatory effects need also to be taken into consideration. Some of these regulatory effects will be common to all members of the functional group. However, other regulatory effects may be relevant for a subset only of the functional group. For instance, zinc sulfate stimulates DAPG production in certain but not all genetic groups of Pseudomonas PGPR strains.

Second, the relationships amongst the different PGPR strains co-occurring in a same rhizosphere are important. Interactions will take place within a functional group, as illustrated above with QS regulation of phenazine production in fluorescent Pseudomonas PGPR. Interactions may also take place between different PGPR functional groups, integrating competitive and inhibitory effects, signal jamming and positive signaling, as well as more indirect processes such as root exudation modifications). These interactions have the potential to modulate spatial colonization patterns of PGPR on roots and to affect PGPR performance. This also suggests that members of different PGPR functional groups can function together, as consortia, with the possibility of synergistic effects or, contrariwise, antagonistic effects. These positive effects may be sought by implementing inoculation procedures in which different types of plant-beneficial microorganisms are used in combination, as highlighted above. Even in this context, interactions between the different microbial strains that are inoculated and indigenous microorganisms (including PGPR) probably matter.

Impact of Plant Genotypes on PGPR Functional Groups

Plants at species, sub-species and variety levels exhibit substantial genetic and phenotypic diversity. In the rhizosphere, different plant genotypes will have a different impact on the number, diversity and activity of microorganisms. This has been shown when comparing different plant species or varieties within species. It entails differences noticeably in root system structure, root exudation profile, and nutrient acquisition. These effects have also been evidenced when considering microbial functional groups of PGPR or where PGPR predominate.

Nitrogen-fixing bacteria are particularly important for plant nitrogen nutrition. The analysis of functional groups indicated that the size and composition of nitrogen-fixing bacteria is influenced by host plant features, both at plant species and variety levels. Analysis of nifH gene transcripts extracted from the rhizosphere showed that only a fraction of the community expresses nifH, and that the corresponding bacterial species differed according to the plant variety, pointing to an influence of plant genotype on the functioning of nitrogen-fixing bacteria. Similar findings were made with the functional group of phosphate solubilizers. Their selection by roots varies according to host plant species.

Other functional groups, such as those involved in plant protection from parasites, act mainly by competition or antagonism, even though direct ISR effects might also take place. For these microorganisms as well, plant genotype can have a major effect on microbial selection processes, as shown with fluorescent pseudomonads producing DAPG or hydrogen cyanide. Plant-genotype differences in rhizosphere ecology may also matter in terms of plant protection efficiency.

Plants have evolved different types of biotic interactions with soil microbial populations, ranging from commensalism to mutualism. Within this continuum of interactions, the plant-PGPR cooperation plays a major role by enhancing growth and health of widely diverse plants. Recent progress has helped to understand key features regarding the modes of action and ecology of plant-PGPR interactions, but major knowledge gaps remain. In terms of molecular signaling and functioning, whether PGPR fine-tune plant hormonal pathways similar to those induced by pathogens and symbionts and trigger yet-unknown specific pathways requires clarification.

Plant growth-promoting rhizobacteria are able to modulate RSA and in fine the vegetative growth and physiology of the whole plant. RSA effects have long been associated with the production of IAA by PGPR. Surprisingly, bacterial modulation of plant auxin distribution and IAA signal transduction pathways, independently of IAA production by PGPR, has also been revealed. It is obviously a step forward in our understanding of plant-PGPR cooperation but it does not fully clarify the bacterial functions and plant hormonal networks involved. Plant hormones regulate genes for the biosynthesis of other hormones or components of hormonal pathways. Consequently, it is possible that PGPR can affect these cross-talks too. It would explain why PGPR can have such pleiotropic effects on plants. One of the major current scientific challenges lying ahead is to understand how these different signaling pathways are integrated to coordinate plant growth and development, and how PGPR influence the plant hormonal network.

Distinct PGPR populations present in a same soil can express plant-beneficial properties in concert. As aforementioned, the relationships between plants and their rhizo-microbiome are complex and vary both according to plant genotypes and soil inhabiting populations (and thus local soil properties, more generally speaking). Next-generation sequencing technologies have started to reveal their taxonomic and functional diversity. They have begun to bring new knowledge on the ecology of PGPR functional groups. In the near future, it is expected that metatranscriptomics and metaproteomics will develop drastically, and will allow further progress on the understanding of the activity and ecological behavior of natural PGPR populations within the rhizosphere. However, given the heterogeneity in space and time of the rhizosphere habitat, samplings at different times and locations within the plant rhizosphere and within fields will be essential to better understand the ecology and performance of PGPR at plant and field plot scales. Nevertheless, despite being very reductionist, mechanistic functional studies using one PGPR and one plant are still useful to investigate the ways PGPR exert beneficial effects on plants. We think that the most important advances on plant-PGPR cooperation will be brought in the future by combining both ecology and functional biology approaches.

Rhizosphere Interactions

Rhizosphere Soil Environment

Plant roots have strong effects on the physical environment in the rhizosphere. Shifts in the physical environment, which are influenced by root manipulation, have significant impacts on the availability and form of substrates used for microbial metabolism.

In turn, this strongly affects the prevalence of different microbial functions, as well as their total biomass within the rhizosphere. This interplay goes both ways, as manipulation by microbes also plays a strong role in the ways that plant roots interact with, benefit from, and influence the soil environment around them.

Water Relations

The availability of water is a major component in determining the diversity and prosperity of soil organisms. Water availability is quantified by the water potential that describes the transport and energy potential state of water in different forms and locations. Plant transpiration, via leaf stomata, is a mechanism for connecting the soil-plant-atmosphere continuum that is driven by a vapor pressure gradient between two physical locations (i.e. the plant root and surrounding soil or the leaf interior and the atmosphere). As water transpires across the leaf surface, tension is created within the leaf that is ultimately transferred down to the root system via adhesion and cohesion of water through the vascular tissues of the plant. This can drastically alter the water potential of the rhizosphere, resulting in large fluctuations in the availability of water and solutes for microbes. This fluctuation is dramatically greater than that of the bulk soil, where changes in water potential are driven predominantly by evaporation that occur at much slower rates. This fluctuation in available water can put pressure on the diversity and total biomass of microbes in the rhizosphere, by selecting for those with better mechanisms for water uptake.

On average, 50% of soils consist as pore space, composed of air and water with differing concentrations of solutes. The constituents of the soil solution provide the substrates necessary for microbes to thrive, ultimately determining the types of biotic and abiotic interactions taking place. Within the rhizosphere, solute activity is significantly higher than that of the bulk soil. One reason is that plant roots secrete ions and organic compounds that interact with organisms and other molecules in the environment around them. Additionally, plant roots uptake solutes in the soil solution, causing changes in ion concentrations that drive the rate and direction of chemical processes within the rhizosphere. As a result, the reinforced microbial population carries out enzymatic processes, such as the transformation of carbon and nitrogen species, at higher rates than in the bulk soil.

Soil Physical Properties

The relative ratio of sand, silt, and clay has large impacts on the availability of water, air, and nutrients for microbial functions. The movement of roots through the soil profile occurs more rapidly in sandy soils than in clayey soils. This is due to sand having larger pore spaces between granules, allowing nutrient and water percolation to occur more rapidly. Increased porosity also allows for plant secreted compounds to extend further away from the root membrane. Therefore, the larger the granule size, the further the associated rhizosphere and microorganisms will extend into the surrounding soil. This means that larger granules require microbes to travel further distances to attain nutrients or engage in interactions with plant roots. This is especially true in soils with very low water potential.

Additionally, soil structure of the rhizosphere is also affected by the increased interactions of microorganisms. In addition to the binding effects of plant mucilages, other root carbon depositions also stimulate soil bacteria biofilm production. These biofilms are composed of extracellular carbon polymers that act as an aggregating mechanism for mineral particles and organic matter. The rate and composition of bacterial exopolymers is highly dependant on the composition of root-secreted compounds, although in general have been shown to be significantly higher in the rhizosphere than in the bulk soil.

Mycorrhizal fungi have also been shown to increase aggregation through the physical constriction of fungal hyphae, as well as through the secretion of their own exopolymer deposits. Further, the sorptive forces of plant roots and mycorrhizae provide an attractive force that promotes mineral alignment and provides a "backbone" for stabilizing aggregate surfaces. In addition, the decomposition of roots and hyphae also increases soil aggregation by the delivery of new organic material, as well as by producing larger pores and preferential flow pathways for air and water. The relationship of roots with associated mycorrhizae can alter both the decomposition rate as well as the final content of the decomposition products.

Soil Chemistry

The effects of pH extend to almost every characteristic of soil that are affected by multiple interactions within the rhizosphere. An example is respiration, a reaction that leads to carbon dioxide (and eventually to bicarbonate/carbonic acid) generation. In addition to the respiration of the roots themselves, the carbon-rich environment of the rhizosphere promotes high levels of respiration by other macro and microorganisms, to a far greater extent than what is occurring in the bulk soil. Increased microbial activity also increases microbial transformations of chemical species. These transformations result in increased acidity, such as nitric acid production during nitrification and sulphuric acid production during sulphur reduction. The influencing effects of pH in the rhizosphere are critical in supporting a biologically diverse microbial community. Manipulation of the pH by plant roots is constantly occurring and, while it is commonly believed that plants manipulate the soil pH in order to change the availability of nutrients, it also undoubtedly influences the abundance and diversity of associated microorganisms.

Microbes are also essential in performing reduction-oxidation (redox) reactions in soil. These redox reactions provide energy for the microbes while also changing the valence of ions in solution. The changes in valence are important for plants, which require essential nutrients to be available, often in inorganic forms in order to be absorbed.

Increases in microbial populations, associated with the rhizosphere, also result in increased redox reactions. Additionally, a study showed that variations in different iron valence concentrations resulted in quantitative differences between rhizosphere microbial populations. While we have been aware of how the rhizosphere promotes increased microbe populations for some time now, we are only beginning to understand how the redox state of various ions are affecting these same populations.

Rhizodeposits and microbial exopolymer secretion also increase the amount of molecular chelation and complexation in the rhizosphere compared to the bulk soil. These complexes provide different functions, depending on the composition. Some of these functions include reducing or increasing solubility, altering the charge, and reducing toxicity of molecules. The changes in solubility are essential for allowing plant roots and mycorrhizae access to highly immobile nutrients, such as phosphate and iron.

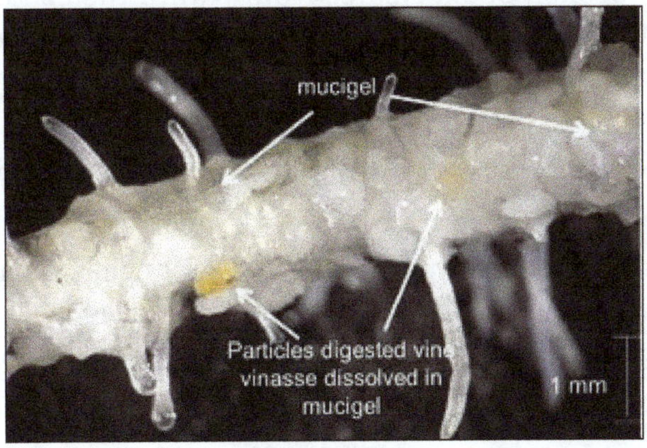

Mucigel.

Plant-derived Compounds

Plant-derived compounds are responsible for providing the additional carbon that allows the rhizosphere to host such a large variety of microorganisms. Plants have mechanisms to control the production and deposition of these compounds into the rhizosphere. The function of these depositions varies vastly, and is generally influenced by the conditions of the soil environment around the root. Additionally, different species of plants appear to favor production of different types of polymers. A study found the rhizosphere populations of canola and potato to contain much more similarities to each other than to strawberry. These plant-derived compounds fall into five predominate categories: exudates, secretions, mucilages, mucigel, and lysates.

Exudates include surplus sugars, amino acids, and aromatics that diffuse out of root cortex cells into the intercellular space (apoplast) and surrounding soil. They comprise the largest category of root secretions and are therefore considered to have the broadest effect in rhizosphere manipulation. The concentration of exudates are magnitudes larger the closer they get to the root, as a result of removal of the compounds by the microbial community faster than the root can replenish. The root controls the rate of exudate release into the rhizosphere through production of solute-specific membrane channels. The best known examples of exudates are organic acids that increase P solubility and reduce Al toxicity. Due to their diffusive nature, exudates are limited to compounds of low molecular weights.

Secretions also known as secondary metabolites, are byproducts of metabolic activity. Because they are released from the cell via active transport, secretions can be of both low and high molecular weight. They are also important in increasing mobilization of insoluble compounds, such as Fe and P. However, the release mechanisms of secretions are less studied than other plant-derived compounds.

Lysates are generally described as all the contents of a cell once the cell is lysed and once the compounds are released into the surrounding soil. When an epidermal root cell dies and is broken open, lysates from within the cell become available to the surrounding microbial community. These components include proteins, nucleic acids, lipids, and various forms of carbohydrates. Lysate compounds are important in maintaining the C/N ratio of soil organic matter. Due to the nature of their release, plants have considerably less control over the concentration of these compounds in soil.

Mucilages are the result of abrasive forces of the root against soil particulate matter that removes cells from the root cap. These newly removed cells consist of cellulose, pectin, starch, and lignin, in which they are often highly recalcitrant and therefore contribute to higher diversity of carbon decomposers.

Mucigel is a slime coating the surface of a root that increases the connectivity between plant roots and the surrounding soil. It is more common on the main body of the root and root hairs than on the tip. During very low soil water potential, mucigel is responsible for allowing plants to continue the uptake water and nutrients, as well as reducing the negative effects of desiccation.

Microbial Communities

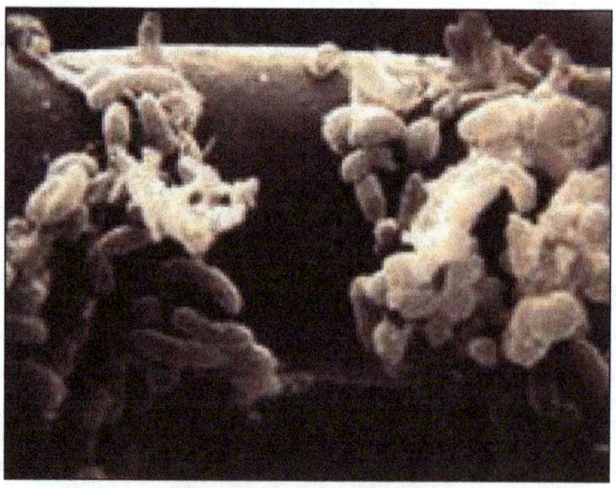

Rhizosphere bacteria.

The rhizosphere can be seen as a competitive environment for a diverse range of microorganisms that inhabit it. Various microorganisms compete with each other for resources such as nutrients, space, and water. Some may even develop more close associations to outcompete their competitors.

Bacteria

Bacteria are the most numerous organisms in the soil. Accordingly, studies suggest that proteobacteria, actinobacteria, non-sporulating rods, pseudomonads, and actinomycetes are among the most dominating populations of bacteria within the rhizosphere. Due to their small mass, they only account for a small amount of the biomass within the soil. Metabolites released from the root hairs of the plants act as chemical signals for bacteria to actively move towards the root surface where they can obtain nutrients. Additionally, bacteria known as plant-growth-promoting-rhizobacteria(PGPR) colonize roots very efficiently. They perform important functions to satisfy plant health and growth in various ways. Furthermore, symbiotic relationships with plants leads to benefits for both the plant and the PGPR, such as better nutrient uptake and stimulation of root growth.

The most common genera of bacteria are the following: Agrobacterium, Mycobacterium, Alcaligenes, Cellulomonas, Micrococcus, and Pseudomonas.

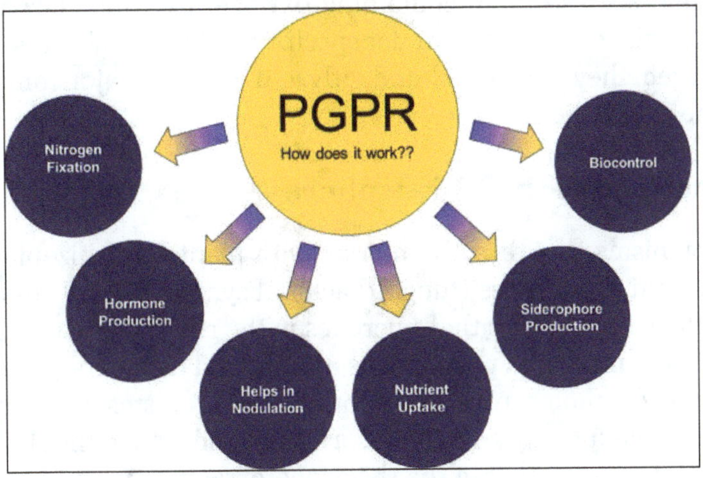

PGPR.

Fungi

Both pathogenic and symbiotic fungi associate with the rhizosphere; however, there are many factors such as plant development and soil type that determine what specific fungal communities live within it. In particular, root exudates have been deemed an important factor when selecting for specific rhizosphere fungi. Zygomycetes and hyphomycetes establish themselves the fastest in the rhizosphere because they only metabolize simple sugars. Pythium and Aphanomyces are pathogenic fungi that cause diseases to plant roots. Furthermore, fungi such as Gibberella form a symbiotic relation with plant roots that stimulates their growth. In times of flooding, when too much moisture accumulates around the base of the tree, a bacterial pathogen named Armillaria may infect and cause root rot.

Archaea

Archaea were initially viewed as extremophiles, but they now have been found in a broad range of environments that include soils, oceans, and even human skin. In comparison to bacteria, there is less known information about the interactions that archaea have with the soil; however, we are certain that they are very important for the process of ammonification within the rhizosphere, which is a process that includes transforming organically bound nitrogen from microbial/plant biomass to ammonia. Furthermore, archaea also play a major role in both the carbon and nitrogen cycle.

Interestingly enough, current research done on archaea within the rhizosphere indicate the extent in which they are involved in rice crops. Much of the global methane gas emissions in the world come from rice paddies that are due largely due in part to the large community of methanogenic archaea producing the methane gas. While soil in rice paddies tend to be anaerobic, oxygen is able to diffuse into roots and create oxic conditions on the root surface and the adjacent rhizosphere soil. Archaea can thrive in oxidized rhizosphere soils due to containing oxygen-insensitive enzymes along with DNA repair mechanisms Therefore, under specific conditions such as areas of high CO_2 and reduced oxygen, they increase abundantly and become major contributors to processes within the rhizosphere.

Biotic Interactions in the Rhizosphere

The primary organisms that directly interact with plants are rhizobial bacteria, fungal mycorrhizae, and endophytes (fungal/bacterial symbionts that reside within plant cells). Some of the main benefits that microbes in the rhizosphere provide to the plant are nitrogen fixation, increased phosphorus uptake, and defense against pathogens. In return, plants provide photosynthetic carbon to the organisms, in which they provide 15-20% to rhizobial nodules, 4-20% to mycorrhizae, and 15% to the rhizoplane. A multitude of interactions are present in the rhizosphere, ranging from beneficial symbiotic relationships to detrimental pathogenic interactions. However, symbiosis can potentially lead to microbes who contribute less than others in maintaining plant health, despite undeservingly gaining the benefits from plants during symbiosis. These organisms are called "cheaters", and there is current research being done on how plants choose their partners ensuring that mutual cooperation will persist. Plants give recognition signals to potential microorganisms for symbiosis; however, these signals are not completely accurate as they can also attract rhizobia and mycorrhizae that can potentially become "cheaters". In a recent study based on the sanctioning hypothesis, soybean plants living in a mutualistic environment with rhizobia were grown in an $Ar:O_2$ atmosphere that was very low on N_2, significantly limiting the rhizobia from fixing nitrogen. As a result, the plant immediately restricted its oxygen influx to rhizobia as a means of punishment, which inhibited further growth of the aerobic bacteria.

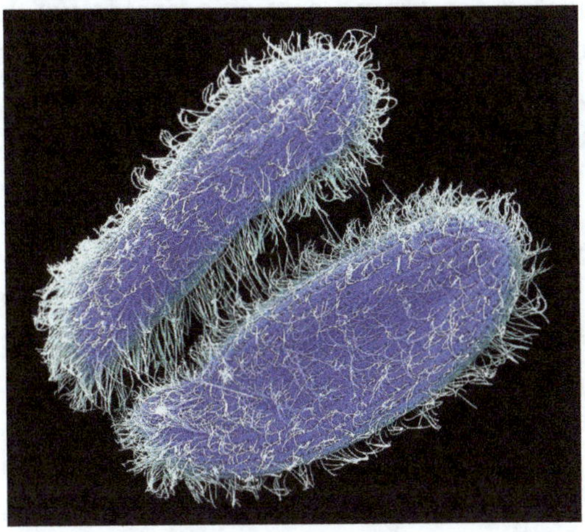

Protozoa.

Impact of Rhizobial Microbes on Plants

Plants secrete secondary metabolites that inhibit certain effective microbes, which can either be beneficial to the plant by pathogen suppression or be detrimental to the plant by competing for nutrients. On the other hand, microbes can be beneficial to the plant by either providing essential macronutrients (Nitrogen, Phosphorus), or by inhibiting the activity of pathogenic microbes; nevertheless, pathogenic microbes can be harmful to the plant by either directly infecting the plant through releasing toxins or by infecting the beneficial microbes around them.

Microorganisms in the rhizosphere contribute chemical and physical modifications that directly affect plants. Chemical changes occur as a result of the conversion of organic matter and microbial biomass to mineral compounds that provides plants with a form of nutrients that are readily available for uptake. In this case, rhizobial bacteria are responsible for reducing manganese (Mn_4^+ to Mn_2^+) and iron (Fe_3^+ to Fe_2^+) to make available for plant uptake. Manganese and iron are vital nutrients that play a tantamount role in lignin synthesis, ultimately increasing resistance from soil-borne plant pathogens from inoculating plant roots. Mycorrhizal fungi are largely responsible for water and nutrient uptake, especially phosphorus, zinc, and copper, and are involved in increasing soil aggregation. In tropical environments where phosphorus is deficient, these fungi are involved in a process called phosphorus fixation, which includes retrieving phosphorus deep in the soil profile where rhizobial bacteria subsequently solubilize the mineral for plant uptake. Furthermore, the fixation of atmospheric dinitrogen by both asymbiotic and symbiotic bacteria results in increases in the available nitrogen pool that can be accessed by plants in the rhizoplane, resulting in the continual release of nutrients to the plant roots.

Physical changes occur primarily through the production of extracellular polymeric substances, such as polysaccharides and glomalin, that improve soil aggregation and soil texture. The presence of mucigel in the rhizoplane is crucial to the water relation of plants since the hydrated compound acts as a protective, lubricative layer for the root cap that aids in the uptake of water. Without this compound, plants would have a higher chance of suffering from times of high water stress since it would be difficult for plants to efficiently penetrate the soil and form a symbiotic relationship with fungi that aid in water uptake.

In addition to the soil texture and organic matter content, the variety, age, and health of the plants play a critical role in both the composition and dynamic of microbial activity in the rhizosphere. This is due to the fact that plant roots cause chemical and physical changes to the soil they inhabit, subsequently affecting the microbial diversity in and around the rhizosphere that select for/against certain microbial populations. Nevertheless, plants can also compete with microorganisms for resources such as water and nutrients, especially if these organisms are not mutually benefiting the plants.

Symbiotic or Mutualistic Relationships

In many symbiotic or mutualistic relationships, plants usually provides a source of carbon to bacteria and fungi. In return, the bacteria or fungus fills some other, more specialized function. In many cases, the association between the two is not absolutely necessary for the survival of both members although it provides significant benefits. An example of a symbiotic relationship is the one between the recently sequenced genome of Laccaria Bicolor (an ectomycorrhizal fungus) and plants such as trees. Laccaria Bicolor shows an excess of enzymes used in ammonia uptake, and a lack of enzymes that would be needed to degrade plant wall material. Thus, it is assumed the fungus is providing nitrogen (and perhaps other nutrients), and a guarantee not to attack the tree in return for glucose. Other than providing plants with nutrients, microbial partners can also help the plant that it forms a relationship with survive in an otherwise inhospitable environment. For example, specific endophytes have been shown to confer heat resistance to grasses that grow in a hydrothermal area, or provide salt tolerance to costal grasses.

Mycorrhizal Fungi

Over 80% of all land plants have a mutualistic relationship with one or more mycorrhizal fungi. Mycorrhizal fungi extend the effective root length by 100 fold or more. Not only do they increase the surface area over which nutrients and water can be taken up by the plant, but they can also reach into smaller pores than can roots. The most valuable asset mycorrhizae provide is accessing immobile nutrients, such as phosphorous. In return, the plant passes a significant amount of its carbon reserves to the fungi. Although, mycorrhizal fungi are beneficial for the most part, they can also be mildly detrimental to the plant, or have circumstances in which the plant feeds from the fungi.

Orchid Mycorrhizal Fungi

An example in which mycorrhizal fungi and plants are beneficial to each other is manifested in the symbiosis they have with orchids. For a small stage in the orchids life, it depends on fungi to provide simple sugars. Orchid seeds cannot germinate unless they have fungi that are attached. In return for its sugars, the orchid allocates trace elements to the fungi. Although this positive interaction is seen, the mycorrhizal fungi do not benefit completely. The plant parasitises the fungus that invades it.

Ectomycorrhizae

Ectomycorrhizae.

Ectomycorrhizae colonize about 3% of plant seeds. These fungi interact with the plant roots so they can obtain sugars since they are poor degraders of cellulose and other plant wall material. The way they interact with the plant root is purely extracellular, in which they produce a net-like structure called a "hartig net" that weaves between the root cortical cells. A sheath or mantle of fungal tissue covering part or all of the root is responsible for most of the increase in surface area caused by mycorrhizae. Some fungi, such as Boletus betulicola, only have a narrow range of plants they can associate with. Other fungi have a much larger range of host plants. One such fungus, Pisolithus tinctorius, associates with 46 tree species and eight genera. Although these fungi generally have a positive symbiosis with plants, some evidence shows that they are becoming invasive to non-native plant species.

Arbuscular Mycorrhizae

Arbuscular mycorrhiza are also referred to as endomycorrhizae because its branched arbuscule grows within the root cortical cell. Direct connection between the plant and fungal cytoplasm allows the mutual transfer of nutrients and carbon between the fungi and the plant to be more efficient. Examples of mycorrhizal fungi include Glomus tenue and Scutellospora. These types of fungi have many benefits on the plant performance and health of the soil. For example, by making up 50% of fungal mycelium in soil, they hold the soil aggregates together. These fungi release a substance called glomalin, a sticky substance that holds the soil together, further increasing its stability; this leads to an more efficient uptake in phosphorus. Because AM create a larger surface area, they can have more access to phosphorus. Because phosphorus is easily taken up, it reduces the amount of fertilizer needed to be applied. This ultimately reduces the amount of eutrophication in the water, therefore decreasing pollutants. Studies have also shown that arbuscular mycorrhiza are beneficial in protecting soil born pathogens, therefore allowing them to grow stronger and healthier.

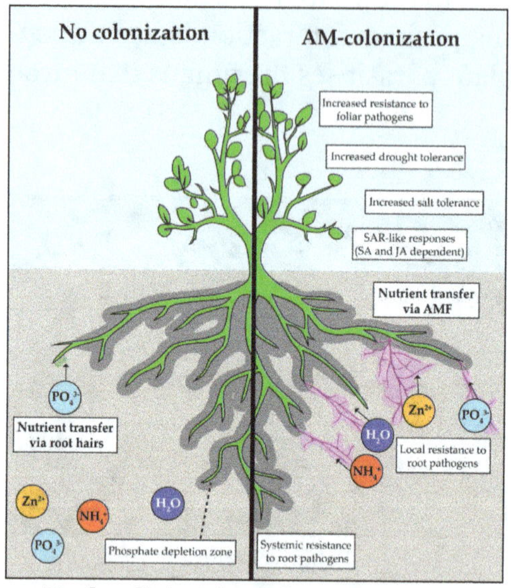

Trade-offs in Arbuscular Mycorrhizae.

Nitrogen Fixing Bacteria

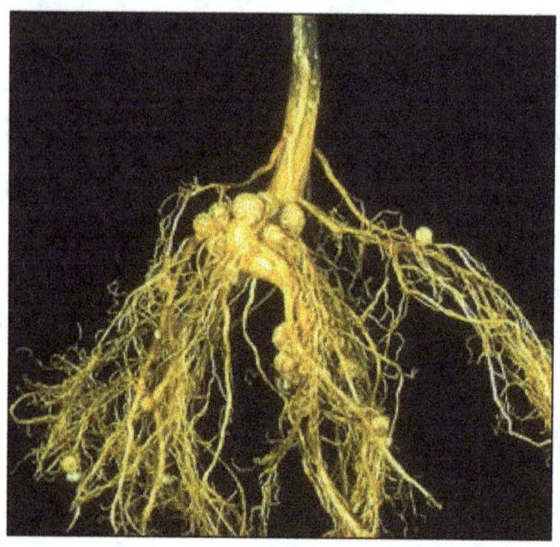

Nodes.

Many organisms have the ability to fix nitrogen although not many actually do. About 86 species in 2 genera of archaea, 38 genera of bacteria, and 20 genera of cyanobacteria can fix nitrogen. This ensures that any nitrogen lost can be replenished. Nitrogen fixation is a very important process in soil, as it benefits the plant in many ways. When the nitrogen is being fixed by bacteria it is directly benefiting the plant by making ammonium that the plant can then utilize to make amino acids and proteins. Another way in which the plant acquires nitrogen is when the nitrogen fixing bacteria dies, and releases an abundant amount of nitrogen into the environment that the plants can then utilize. Nitrogen fixers not only benefit their symbiont, but can also benefit other plants that are not involved in the symbiotic relationship. Some nitrogen can be leaked and taken up by other plants. Most of the nitrogen is recycled into the soil as legume plants decompose and die. Due to the high energetic cost of fixing dinitrogen, a significant part of nitrogen fixation occurs near the plant roots, where there is an influx of sugar to power the process. Some nitrogen fixation occurs in the rhizosphere by free-living bacteria (the ability is found only in prokaryotes). When these otherwise free living bacteria form a close association with a plant they're called "associative symbiotes". An example of a free living nitrogen fixer is Bradyrhizobium.

There are several drawbacks to this lifestyle. For example, Nitrogen fixing bacteria are sensitive to fluctuations in oxygen. Although many are aerobes who use oxygen as a terminal electron acceptor, some oxygen can poison the nitrogenase enzyme. Free-living bacteria need to take energetically costly steps such as having extremely fast respiration or producing large amounts of slime to protect nitrogenase. Another disadvantage associative symbiotes have to deal with is direct competition with the rest of the rhizosphere for the available carbon. Experiments indicate that carbon is a limiting factor in rhizosphere nitrogen fixation.

An alternate form of nitrogen-fixing lifestyle is to actually live in the plant root cells themselves. Bacteria who do this are called "mutualistic symbionts." An example of one such bacteria is Rhizobia. These bacteria form nodules on plants, which are huge masses of bacteria living inside plant cells that have been modified for this purpose. The process of forming a nodule requires recognition between bacteria and plant (usually very specific), and gene expression changes in both players. The plant provides carbon to the bacteria and the bacteria provide fixed nitrogen to the plant. The masses of bacteria produce leghemoglobin to protect nitrogenase from oxygen. Leghemoglobin gives the nodules a pink color, which is an indication that nitrogen fixation is occurring. If the nodule is white or gray in color, the nodule is fixing very small amounts of nitrogen or none at all. Much like the hemoglobin in the human body, leghemoglobin binds to oxygen to its subunits and holds onto it so that it cannot be used.

Nodules containing leghemoglobin.

Legume Nodules

The small nodules are formed by rhizobia bacteria entering the cell root making nodules. They are visible after 2-3 weeks of growing. These are the type of nodules in which a pink color is given by the leghemoglobin.

Perennial Legume Nodules

These long finger-like nodules are half an inch in diameter and can found on alfalfa and clover. When these nodules are mature, they have a larger center with many fingers like protrusion. These nodules grow mostly around the tap root and with live for quite a long time.

Perennial nodules.

Annual Legume Nodules

These spherical pea-sized nodules can be found on beans, peanuts, and soybeans. Unlike the perennial nodule, these nodules are short-lived. When plants are filling their pods, the nodules will not fix nitrogen. This is because the plant is too busy giving its nutrients to the seeds instead of the nodule.

Frankia

Frankia interactions.

Frankia are actinomycete filamentous bacterium that form nodules. These nodules inhabit the roots of actinorhizal plants such and shrubs and trees. The process used for Intracellular infection by Frankia is quite complex. Frankia and plants use specific signals that activate a symbiotic signaling pathway. Research today is not clear what the biochemical signal characterizations are of Frankia, but have found that Flavonoids improve nodulation and are necessary for the infection. Recent genome sequencing has allowed scientists to theorize how and why prokaryotes and plant symbioses occurred, adaptations, metabolic diversity, and horizontal gene flow among symbiotic prokaryotes.

Inoculants

Inoculation of fields with the soil from other fields has been practiced for centuries as a method of improving agricultural productivity. Modern seed companies are dipping seeds in beneficial inoculants before sale and many commercial microbial inoculants are now sold on the consumer market. In addition, organic food and farming is becoming increasingly popular and use of beneficial microbial and fungal inoculants has increased substantially. However, the specific conditions surrounding the seed at germination and specific soil characteristics determine the ability of the soil to support a population of desirable rhizosphere microbes. In addition, there are still concerns over the uncertain effects of transplanting microorganisms, including the effects it may have on native ecosystems. Currently, the primary organisms being used for these products are mycorrhizal fungi, dinitrogen-fixing bacteria, and beneficial rhizobacteria.

Root Microbiome

The root microbiome (also called rhizosphere microbiome) is the dynamic community of microorganisms associated with plant roots. Because they are rich in a variety of carbon compounds, plant roots provide unique environments for a diverse assemblage of soil microorganisms, including bacteria, fungi and archaea. The microbial communities inside the root and in the rhizosphere are distinct from each other, and from the microbial communities of bulk soil, although there is some overlap in species composition.

Different microorganisms, both beneficial and harmful affect development and physiology of plants. Beneficial microorganisms include bacteria that fix nitrogen, promote plant growth, mycorrhizal fungi, mycoparasitic fungi, protozoa and certain biocontrol microorganisms. Pathogenic microorganisms also span certain bacteria, pathogenic fungi and certain nematodes that can colonize the rhizosphere. Pathogens are able to compete with protective microbes and break through innate plant defense mechanisms. Apart from microbes that cause plant diseases, certain bacteria that are pathogenic and can be carried over to humans, such as *Salmonella*, enterohaemorhagic *Escherichia coli*, *Burkholedria (ceno)cepacia*, *Pseudomonas aeruginosa*, and *Stenotrophomonas maltophilia* can also be detected in root associated microbiome and in plant tissues.

Root microbiota affect plant host fitness and productivity in a variety of ways. Members of the root microbiome benefit from plant sugars or other carbon rich molecules. Individual members of the root microbiome may behave differently in association with different plant hosts, or may change the nature of their interaction (along the mutualist-parasite continuum) within a single host as environmental conditions or host health change.

Despite the potential importance of the root microbiome for plants and ecosystems, our understanding of how root microbial communities are assembled is in its infancy. This is in part because until recent advances in sequencing technologies, root microbes were difficult to study due to high species diversity, the large number of cryptic species, and the fact that most species have yet to be retrieved in culture. Evidence suggests both biotic (such as host identity and plant neighbor) and abiotic (such as soil structure and nutrient availability) factors affect community composition.

Types of Symbioses

Root associated microbes include fungi, bacteria, and archaea. In addition, other organisms such as viruses, algae, protozoa, nematodes and arthropods are part of root microbiota. Symbionts associated with plant roots subsist off of photosynthetic products (carbon rich molecules) from the plant host and can exist anywhere on the mutualist/parasite continuum.

Root symbionts may improve their host's access to nutrients, produce plant-growth regulators, improve environmental stress tolerance of their host, induce host defenses and systemic resistance against pests or pathogens, or be pathogenic. Parasites consume carbon from the plant without providing any benefit, or providing too little benefit relative to what they cost in carbon, thereby compromising host fitness. Symbionts may be biotrophic (subsisting off of living tissue) or necrotrophic (subsisting off of dead tissue).

Mutualist-parasite Continuum

While some microbes may be purely mutualistic or parasitic, many may behave one way or the other depending on the host species with which it is associated, environmental conditions, and host health. A host's immune response controls symbiont infection and growth rates. If a host's immune response is not able to control a particular microbial species, or if host immunity is compromised, the microbe-plant relationship will likely reside somewhere nearer the parasitic side of the mutualist-parasite continuum. Similarly, high nutrients can push some microbes into parasitic behavior, encouraging unchecked growth at a time when symbionts are no longer needed to aid with nutrient acquisition.

Composition

Roots are colonized by fungi, bacteria and archaea. Because they are multicellular, fungi can extend hyphae from nutrient exchange organs within host cells into the surrounding rhizosphere and bulk soil. Fungi that extend beyond the root surface and engage in nutrient-carbon exchange with the plant host are commonly considered to be mycorrhizal, but external hyphae can also include other endophytic fungi. Mycorrhizal fungi can extend a great distance into bulk soil, thereby increasing the root system's reach and surface area, enabling mycorrhizal fungi to acquire a large percentage of its host plant's nutrients. In some ecosystems, up to 80% of plant nitrogen and 90% of plant phosphorus is acquired by mycorrhizal fungi. In return, plants may allocate ~20-40% of their carbon to mycorrhizae.

Fungi

Mycorrhizae

Mycorrhizal defines symbiotic interaction between plants and fungus. Fungi are important to decompose and recycle organic material, however the boundaries between pathogenic and symbiotic lifestyles of fungi are not always clear-cut. Most of the time the association is symbiotic with fungus improving acquisition of nutrients and water from soil or increasing stress tolerance and fungus benefiting from carbohydrates produced by plant. Mycorrhizae include a broad variety of root-fungi interactions characterized by mode of colonization. Essentially all plants form mycorrhizal associations, and there is evidence that some mycorrhizae transport carbon and other nutrients not just from soil to plant, but also between different plants in a landscape. The main groups include ectomycorrhizae, arbuscular mycorhizae, ericoid mycorrhizae, orchid mycorrhizae, and monotropoid mycorrhizae. Monotropoid mycorrhizae are associated with plants in the monotropaceae, which lack chlorophyll. Many Orchids are also achlorophyllous for at least part of their life cycle. Thus these mycorrhizal-plant relationships are unique because the fungus provides the host with carbon as well as other nutrients, often by parasitizing other plants. Achlorophyllous plants forming these types of mycorrhizal associations are called mycoheterotrophs.

Endophytes

Endophytes grow inside the plant tissues - roots, stems, leaves - mostly symptomless, however when plant ages they can become slightly pathogenic. They may colonize inter-cellular spaces, the root cells themselves, or both. Rhizobia and dark septate endophytes (which produce melanin, an antioxidant that may provide resilience against a variety of environmental stresses) are famous examples.

Bacteria

Zone of soil surrounding the roots is ritch in nutrients released by plants and is therefore attractive growth medium for both beneficial and pathogenic bacteria. Root associated beneficial bacteria promote plant growth and provide protection from pathogens. They are mostly rhizobacteria that belong to Proteobacteria and Firmicutes, with many examples from Pseudomonas and Bacillus genera. Rhizobium species colonize legume roots forming nodule structures. In response to root exudates, rhizobia produce Nod signalling factors that are recognized by legumes and induce formation of nodules on plant roots. Within these structures Rhizobium perform nitrogen fixation that proves plant with nitrogen source. In turn, plants provide bacteria with carbon source to energize the nitrogen fixation. In addition to nitrogen fixation, Azospirillum species promote plant growth through production of growth phytohormones (auxins, cytokinins, gibberellins). Due to these phytohormones root hairs expand to occupy larger area and better acquire water and nutrients. Pathogenic bacteria that infect plants infect plant roots are most commonly from Pectobacterium, Ralstonia, Dickeya and Agrobacterium genera. Among the most notorious are Pectobacterium carotovorum, Pectobacterium atrosepticum, Ralstonia solanacearum, Dickeya dadanthi, Dickeya solani and Agrobacterium tumefaciens.

Bacteria attach to roots in a certain biphasic mechanism with two steps – first weak, non-specific binding and then switch to second – strong irreversible residence phase. Both beneficial and pathogenic bacteria attach in this fashion. Bacteria can stay attached to the outer surface or, certain endophytes or pathogens can colonize the inner root. Primary attachment is governed by chemical forces or certain extracellular structures such as pili or flagella. Secondary attachment is mainly characterized by synthesis of cellulose, extracellular fibrils and specific attachment factors such as surface proteins that help bacteria aggregate and form colonies.

Archaea

Though archaea are traditionally thought of as extremophiles, microbes belonging to extreme environments, advances in metagenomics and gene sequencing reveal that archaea are ubiquitous, found in nearly any environment including the root microbiome. For example, root-colonizing archaea have been discovered in maize, rice, wheat, and mangroves. Methanogen and ammonium-oxidizing archaea are prevalent members of the root microbiome, especially in anaerobic soils and wetlands. Archaeal phylums found in the root microbiome include Euryarchaeota, Thaumarchaeota, and Crenarchaeota.

The presence and relative abundance of archaea in various environments suggest that they likely play an important role in the root microbiome. Archaea have been found to promote plant growth and development, provide stress tolerance, improve nutrient uptake and protect against pathogens. For example, *Arabidopsis thaliana* colonized with an ammonia-oxidizing soil archaea, *Nitrosocosmicus oleophilius,* exhibited increased shoot weight, photosynthetic activity and immune response.

Examination of microbial communities in soil and roots identify archaeal organisms and genes that occupy functions similar to that of to bacteria and fungi, such as auxin synthesis, protection against abiotic stress and nitrogen fixation. In some cases, key genes for plant growth and development, such as metabolism and cell wall synthesis, are more prevalent in archaea than bacteria.

Archaeal presence in the root microbiome can also be affected by plant hosts, which can change the diversity, presence and health of archaeal communities.

Viruses

Viruses also infect plants via the roots, however to penetrate the root tissues they typically use vectors such as nematodes or fungi.

Assembly Mechanisms

There is an ongoing debate regarding what mechanisms are responsible for assembling individual microbes into communities. There are two primary competing hypotheses. One is that "everything is everywhere, but the environment selects," meaning biotic and abiotic factors pose the only constraints, through natural selection, to which microbes colonize what environments. This is called the niche hypothesis, and its counterpart is the hypothesis that neutral processes, such as distance and geographic barriers to dispersal, control microbial community assembly when taxa are equally fit within an environment. In this hypothesis, differences between individual taxa in modes and reach of dispersal explain the differences in microbial communities of different environments. Most likely, both natural selection and neutral processes affect microbial community assembly, though certain microbial taxa may be more restricted by one process or the other depending on their physiological restrictions and mode of dispersion.

Microbial dispersal mechanisms include wind, water, and hitchhiking on more mobile macrobes. Microbial dispersion is difficult to study, and little is known about its effect on microbial community assembly relative to the effect of abiotic and biotic assembly mechanisms, particularly in roots.

The taxa within root microbial communities seem to be drawn from the surrounding soil, though the relative abundance of various taxa may differ greatly from those found in bulk soil due to unique niches in the root and rhizosphere.

Biotic Assembly Mechanisms

Different parts of the root are associated with different microbial communities. For example, fine roots, root tips, and the main root are all associated with different communities, and the rhizosphere, root surface, and root tissue are all associated with different communities, likely due to the unique chemistry and nutrient status of each of these regions. Additionally different plant species, and even different cultivars, harbor different microbial communities, probably due to host specific immune responses and differences in carbon root exudates. Host age affects root microbial community composition, likely for similar reasons as host identity. The identity of neighboring vegetation has also been shown to impact a host plant's root microbial community composition.

Abiotic Assembly Mechanisms

Abiotic mechanisms also affect root microbial community assembly because individual taxa have different optima along various environmental gradients, such as nutrient concentrations, pH, moisture, temperature, etc. In addition to chemical and climatic factors, soil structure and disturbance impact root biotic assembly.

Rhizosphere Microorganisms

The rhizosphere is the narrow region of soil system that is immediately near to the root surface and that is more affected by rhizospheric phenomena. The term "rhizosphere" was first time used by Lorenz Hiltner in 1904. The root exudates include carbohydrates, sugar, organic acids, vitamins, flavonoids, nucleotides, enzymes, hormones, and volatile compounds, inorganic ions, and gaseous molecules. The exudates act as messengers that stimulate interactions between roots and soil organisms.

Therefore, rhizosphere is the most dynamic environment in the soil with a high microbial diversity. It is well established that microorganism's population is higher in rhizosphere than bulk soil.

In the rhizosphere, the populations of microorganisms may fluctuate from few thousands to millions. This plant-microbe interaction has shown many beneficial effects on plant growth and developments, PGP increased availability and uptakes of nutrients, and disease suppression. However, increased resistance to abiotic and biotic stresses, all these leads to increases in crop productivity and soil sustainability as well as nutrient availability.

Table: Presence of microbes in the rhizosphere and non-rhizosphere soil system.

Microorganisms	Rhizosphere	Non-rhizosphere	R:S Ratio
	Microbes g^{-1} Dry Soil		
Bacteria	$1200*10^6$	$53*10^6$	23.0
Fungi	$12*10^5$	$1*10^5$	12.0
Actinomycetes	$46*10^6$	$7*10^6$	7.0
Algae	$5*10^3$	$27*10^3$	0.2

Table: Microbial-mediated soil nutrient availability.

Nutrient	Microbial transformation.
Sulfur	Mineralization, immobilization, oxidation, reduction.
Nitrogen	Immobilization, nitrification, denitrification, urea hydrolysis, N_2 fixation.
Phosphorus	Mineralization, immobilization, extracellular P activity, facilitated uptake by VAM.
Potassium	K solubilization, mobilization, and mineralization.
Zinc	Facilitated uptake by mycorrhizal fungi, Zn solubilization.
Copper	Facilitated uptake by exudates and mycorrhizal fungi.
Iron	Change in oxidation state, production of siderophores, chelation.
Managanese	Change in oxidation state in soil system.

Role of Rhizosphere Microorganisms in Soil Sustainability and Nutrient Acquisition

Rhizosphere microorganisms play important roles in agriculture mainly by improving soil-plant nutrition and health, as well as soil quality of the ecosystem. The rhizosphere is a multifaceted, complex "melting pot" of components and processes affecting plant growth and development and thus ecosystem functioning. The plant-microbe interactions may be classified into beneficial, neutral, or harmful to the plant, depending on the beneficial and harmful effect on specific microorganisms and plants. Exploring this multicomponent complex of rhizosphere and its possible relationship to soil sustainability will become crucial in order to maintain food production with reduced fertilizers and pesticide inputs.

Organic Matter (OM) Decomposition

The OM has been considered as one of the most important components in improving soil health and agricultural sustainability. The rhizosphere microorganisms play an important role in regulating soil organic matter (SOM) decomposition and nutrient cycling due to the direct influences by roots that are often referred as rhizosphere processes. These rhizosphere processes may include exudates, water uptake, nutrient mobilization by roots and microorganisms, and SOM decomposition. These efficient rhizosphere processes are major paths for nutrient and water availability.

Therefore, rhizosphere effect on the SOM decomposition is significant in mediating plant-soil interactions.

SOM decomposition is mediated by rhizosphere processes. This OM in the soil system is more or less continuously transformed by different chemical and biochemical mechanisms into various ranges of compounds. The SOM contains various compounds which are the main energy source for the soil biota. Soil organisms have appropriate biochemistries and life processes to embezzle such energy for their growth; in doing so, the compounds are further transformed and cycled between compartments.

Nutrient Transformation and Availability

Plant-microbe-soil interactions strongly modulate nutrient availability in soil system.

In rhizosphere soil, the activity of microbes and plant roots is clearly distinct from the bulk soil. The plant/crop roots must compete with the root system of neighboring plant species for space, water, and mineral nutrients.

Next to water and temperature, nutrients are the most crucial environmental factors required for growth and development of crops. Soil-plant-microbe interactions have various effects such as positive, negative, and neutral; these efficient systems help to remediate the polluted soil system.

These various efficient rhizosphere microorganisms like free-living and symbiotic nitrogen-fixing rhizobacteria contribute a lot to meet nutrient demand.

Moreover, the phosphate and potassium solubilizing bacteria (PSB and KSB) have the capability to convert inorganic unavailable P and K form to soluble forms available to plants. The soil biota is also involved in the cycling of other nutrient elements of pertinence to agricultural production systems including various macro- and micronutrients. The mycorrhizae (VAM) can also influence the uptake of a range of elements by plants, including macro- and microelements, its work mainly phytoremediation or uptakes of heavy metals.

Other than rhizobacteria, fungi are fundamental for soil ecosystem functioning especially in agricultural soils. It was estimated that the 1.5 million fungal species are present in the natural ecosystems but only few (5–10%) have been described formally. The arbuscular mycorrhizal fungi (AMF) increase the plant growth. The AMF have a high affinity toward enhanced P nutrition in plants. These efficient AMF could also scavenge the available P, Fe, and Zn through their extra radical hyphae. Additional to their contribution in plant nutrition, AMF are able to improve soil structure and mitigate plant stress.

These efficient rhizosphere microbes have shown their influence toward the uptake of various macro- and micronutrients. In another field study, inoculation of chickpea and barley with strains of M. mediterraneum significantly enhanced the K, Ca, and Mg along with P and N in both crop plants.

Table: Microbiological properties in bulk and rhizosphere soils of different plantations.

Tree Species		MBC	MBN	MBP	Microbial C/N
	Mg kg^{-1}				
Siberian elm	Bulk soil	36.8(4.9)	7.0(0.6)	1.8(0.1)	5.2(0.8)
	Rhizosphere soil	52.7(5.2)	9.4(1.0)	2.1(0.2)	5.6(1.0)
	T-test	**	*	NS	NS

Simon poplar	Bulk soil	73.6(7.2)	6.7(0.2)	2.6(0.1)	10.9(1.2)
	Rhizosphere soil	90.7(6.4)	8.2(0.6)	2.8(0.2)	11.1(1.1)
	T-test	*	*	NS	NS
Mongolian	Bulk soil	45.6(3.2)	4.9(0.5)	1.6(0.1)	9.4(1.0)
	Rhizosphere soil	48.4(4.2)	8.0(0.9)	1.8(0.2)	6.1(0.7)
	T-test	NS	**	NS	*

- Data are means with standard errors given in parentheses (n = 5). NS denotes no significant difference (by paired samples T-test).
- Significant differences between bulk and rhizosphere soils at P < 0.05.
- ** Significant differences between bulk and rhizosphere soils at P < 0.01.

Plant Growth-promoting (PGP) Activities

The bacteria present in the rhizosphere act as PGPR; they have the ability to enhance PGP atmospheric N fixation, hormone production, specific enzymatic activity, and plant and insect protection by producing antibiotic and other pathogen-depressing substances. However, the N_2-fixing, PSB, and KSB are important for crop plants as they increase N, P, and K uptake and play a crucial role as PGPR in the biofertilization. In general, huge bulk inorganic/ chemical fertilization is applied to replenish soil N and P demand with the resultant in high cost and environmental risk. Most of the P-fixed in the form of insoluble compounds is unavailable to plants. Thus, the application of such efficient microbes as eco-friendly biological agents may contribute to minimize the use of expensive P and K fertilizers.

Biocontrol Agents

A broad range of efficient microorganisms carry out biocontrol activities in the rhizosphere soils. These microbes play a major role to control diseases, which is a form of biological control and is an environment-friendly approach. Many rhizosphere bacteria have been reported to produce antifungal compounds like HCN, phenazines, pyrrolnitrin, 2, 4-diacetylphloroglucinol, and pyoluteorin. The biocontrol ability of the PGPR depends on a wide variety of traits, such as the production of various antibiotic compounds, iron chelators, and exoenzymes.

Table: Effect of PGPR on various PGP activities under sustainable agriculture.

PGPRs	Responses	Crop
Pseudomonas and Azotobacter	Stimulates growth and Yield.	Chickpea
P. putida strain R-168	Enhance seed germination, growth parameters.	Maize

P. fluorescens strain R-93, P. fluorescens DSM 50090, P. putida DSM291, and A. lipoferum DSM 1691	Increase growth, leaf nutrient contents and yield.	Banana
P. putida strains R-168 and DSM-291	Improves seed germination, seedling growth and yield.	Maize
R. leguminosarum and Pseudomonas sp.	Enhance yield and P Uptake.	Wheat
P. putida strain	Growth stimulation.	Tomato
Azospirillum brasilense	Improve growth Attributes.	Wheat and maize
P. fluorescens strain	Growth and yield.	Pearl millet
Azotobacter and Azospirillum strains	Growth and productivity.	Canola
P. alcaligenes PsA15 and B. polymyxa	Enhance nutrient uptake.	Maize
Rhizobium leguminosarum	Direct growth promotion.	Canola and lettuce

Soil Bioremediations

The soil is an important habitat for thousands of microbes; these efficient microbes are habitats between soil particles and soil organic matter in rhizosphere. Many soil microbes are capable of degrading toxic compounds including herbicides, pesticides, solvents, and other organic compounds and might provide a reasonable and effective measure of disposing toxic compounds.

Due to the degrading and sequestration capability of the microbial communities toward the toxic compounds and heavy metals (HMs), microbes have been used for bioremediation of soil contaminated with HMs.

The organic pollutants can be degraded by plants/crops/microbes through biochemical reactions taking place within the rhizosphere. The phytodegradation of organic pollutants may also be enhanced by rhizospheric activity. The AMF were also demonstrated to PGP and degradation of toxic pollutants in soils with increased pollutant concentrations especially ectomycorrhizal fungi.

They promote the establishment of plants at HM-contaminated soil by immobilizing heavy metals in the soil, thereby reducing the availability of metals to plants.

Table: PGPR as biocontrol agents against various plant diseases.

PGPRs	Experimental condition	Disease
Bacillus strain	Greenhouse	Blight of squash
B. pumilus strain SE34	Laboratory	Blue mold disease in tobacco
B. pumilus strain INR7	Field	Bacterial wilt in cucumber
B. pumilus strain SE34 and B. subtilis strain IN937b	Greenhouse	Cucumber mosaic cucumovirus (CMV) in tomato
B. cereus strains B101R, B212R, and A068R	Greenhouse	Foliar diseases in tomato

Pseudomonas spp.	Rice	Rice sheath rot
P. fluorescens	Saline field	Saline resistance in groundnut
P. fluorescens	Rice	Sheath blight disease
B. subtilis strain GBO3 and B. pumilus strain INR7	Greenhouse/field	Downy mildew in pearl millet
Bacillus sp. and Azospirillum strains SPS2	Greenhouse	Rice blast
B. amyloliquefaciens 937a and B. subtilis 937b	Field	Tomato mottle virus
P. fluorescens strain CHAO + chitin bio-formulations	Greenhouse/field	Banana bunchy top virus (BBTV)
B. strains BB11 and FH17	Greenhouse	Blight of bell pepper
Burkholderia strains MBf21 and MBf15	In vitro/in vivo	Maize rot

Drought and Nutrient Stress/Deficiency

Diverse environmental stresses like drought, salinity, and nutrient deficiencies are connected to soil-derived limitations. These stresses are inducing a series of morphological, physiological, biochemical, and molecular changes in the plant root system to adapt to the stress conditions and to modify the environment from which the stress is originated. Variation in the soil nutrient composition is a common condition in agricultural systems. Indeed, only ~30 to 50% of applied N fertilizers and ~10 to 20% of P fertilizers are used in crops.

Furthermore, there is a general trend worldwide in reducing the use of chemical inputs in agriculture due to environmental concerns. Water and nutrient acquisition from soils is governed by plant roots and their interaction with the abiotic and biotic components of soil. The soil biotic component includes microorganisms that are living symbiotically or in association with plant roots. These microorganisms might enhance acquisition of nutrients in low-nutrient soils and alleviate the effects of low water availability. But there are only few studies that provided information regarding rhizosphere soil microorganisms to plants in mitigating with stresses derived from nutrient and water availability.

Table: Soil bioremediation of pollutants using plant-microbe interaction.

Plant/crop	Pollutants	Microorganisms
Populus deltoides niger	1,4-Dioxane	Actinomycetes
Senecio glaucus	Oil	Arthrobacter
Barley	2, 4-D	Burkholderia cepacia
Populus deltoides	PAH	Kurthia sp.
		Micrococcus sp.

Alfalfa	Pyrene, anthracene, and phenanthrene	Not identified
Oat, lupine, and rape	Pyrene	Not identified
Prairie grasses	PAHs	Not identified
Grasses	Naphthalene	Pseudomonas putida
Wheat	2, 4-D	Pseudomonas putida
Sugar beet	PCBs	Pseudomonas fluorescens

- PAHs: Polyaromatic hydrocarbons.
- PCBs:Ppolychlorinated biphenyls.
- 2,4-D: 2,4-dichlorophenoxyacetic acid.

Management of Rhizosphere System for Soil Sustainability and Productivity

It is well known that nutrient has unsounded effects on many aspects of root development, thus accompanying rhizosphere process. Most of the mineral nutrients were taken up by plants through the rhizosphere where root exudates play a dominant role in driving interactions among soil-plant-microbe system. However, the plant root exudates consist of a complex mixture of various acids, sugars, and enzymes, which have major direct or indirect effects on the acquisition of mineral nutrients required for PGP activities. The rhizospheric chemistry/phenomena can enhance P mobilization and acquisition from soil by plants. It is suggested that optimizing crop combination and BMPs, RMPs, and INM in intercropping systems through better understanding of rhizosphere interactions provides an effective approach to improve sustainable crop production with high yield, significantly higher NUE.

References

- Rhizosphere, agricultural-and-biological-sciences: sciencedirect.com, Retrieved 15 May, 2020
- Bloemberg, Guido V.; Lugtenberg, Ben J. J. (August 2001). "Molecular basis of plant growth promotion and biocontrol by rhizobacteria" (PDF). Current Opinion in Plant Biology. 4 (4): 343–350. doi:10.1016/S1369-5266(00)00183-7. PMID 11418345
- Rhizosphere-Interactions: microbewiki.kenyon.edu, Retrieved 19 July, 2020
- Nguyen C (2009), "Rhizodeposition of Organic C by Plant: Mechanisms and Controls", Sustainable Agriculture, pp. 97–123, doi:10.1007/978-90-481-2666-8_9, ISBN 978-90-481-2665-1
- Rhizosphere-Microorganisms-Towards-Soil-Sustainability-and-Nutrient-Acquisition-319923182: researchgate.net, Retrieved 26 August, 2020
- Plant genotype strongly modifies the structure and growth of maize rhizosphere microbial communities. Soil Biol. Biochem. 42 2276–2281 10.1016/j.soilbio.2010.08.029

PERMISSIONS

All chapters in this book are published with permission under the Creative Commons Attribution Share Alike License or equivalent. Every chapter published in this book has been scrutinized by our experts. Their significance has been extensively debated. The topics covered herein carry significant information for a comprehensive understanding. They may even be implemented as practical applications or may be referred to as a beginning point for further studies.

We would like to thank the editorial team for lending their expertise to make the book truly unique. They have played a crucial role in the development of this book. Without their invaluable contributions this book wouldn't have been possible. They have made vital efforts to compile up to date information on the varied aspects of this subject to make this book a valuable addition to the collection of many professionals and students.

This book was conceptualized with the vision of imparting up-to-date and integrated information in this field. To ensure the same, a matchless editorial board was set up. Every individual on the board went through rigorous rounds of assessment to prove their worth. After which they invested a large part of their time researching and compiling the most relevant data for our readers.

The editorial board has been involved in producing this book since its inception. They have spent rigorous hours researching and exploring the diverse topics which have resulted in the successful publishing of this book. They have passed on their knowledge of decades through this book. To expedite this challenging task, the publisher supported the team at every step. A small team of assistant editors was also appointed to further simplify the editing procedure and attain best results for the readers.

Apart from the editorial board, the designing team has also invested a significant amount of their time in understanding the subject and creating the most relevant covers. They scrutinized every image to scout for the most suitable representation of the subject and create an appropriate cover for the book.

The publishing team has been an ardent support to the editorial, designing and production team. Their endless efforts to recruit the best for this project, has resulted in the accomplishment of this book. They are a veteran in the field of academics and their pool of knowledge is as vast as their experience in printing. Their expertise and guidance has proved useful at every step. Their uncompromising quality standards have made this book an exceptional effort. Their encouragement from time to time has been an inspiration for everyone.

The publisher and the editorial board hope that this book will prove to be a valuable piece of knowledge for students, practitioners and scholars across the globe.

INDEX

www.ingramcontent.com/pod-product-compliance
Lightning Source LLC
LaVergne TN
LVHW080628210926
848485LV00003B/231

* 9 7 8 1 6 3 9 8 9 4 4 2 0 *